Soils, Earthwork, and Foundations

A Practical Approach, Based on 2012 IRC® and IBC®

Kirby T. Meyer, P.E., D.GE, F.ASCE

INTERNATIONAL CODE COUNCIL®

Soils, Earthwork, and Foundations:
A Practical Approach
Second Edition

ISBN: 978-1-60983-448-7

Cover Design:	Duane Acoba
Publications Manager:	Mary Lou Luif
Project Editor:	Greg Dickson
Layout:	Amy O'Farrell

COPYRIGHT © 2013

INTERNATIONAL
CODE COUNCIL®

First Printing: January 2013

PRINTED IN THE U.S.A.

39531T017157

PREFACE to 2nd EDITION

The subjects in the first edition of this book were soils, earthwork, and foundations with insights into design, plan review, construction, and inspection. That book and the 2nd edition are intended to fill a void in technical publishing that has long existed between textbooks intended for use by professional engineers and knowledge gained by experience and on the job training that has been the traditional training ground for inspectors, construction personnel, and technicians. The contents are based on sound principles and up-to-date technology sharpened by the author's extensive construction exposure and forensic studies of failures. The first edition contained frequent references and coordination with the applicable chapters of the 2006 *International Residential Code*® and 2006 *International Building Code*®. The 2nd edition makes similar references to the 2012 *International Residential Code*® (IRC®) and the 2012 *International Building Code*® (IBC®). This is a unique feature rarely found in any similar textbook on the subject.

Studying the book will not make the reader an expert professional but will permit the reader to better communicate with geotechnical and structural professionals and benefit those who do code enforcement, plan reviews, construction inspections, laboratory testing, and construction related to foundations and earthwork. Architects and engineers will also benefit from the information on design and field applications. Engineering students and graduate engineers who are starting to work as professionals will find it helpful in bringing together coursework topics and practical applications.

While it is labeled a practical book, many of the topics are not thoroughly understood by all practicing engineers. One difficulty is that geotechnical engineers understand their discipline very well but may not always be familiar with the work of structural engineers; the opposite is also true. Designing foundations requires some crossover knowledge of geotechnical engineering, and geotechnical engineers should understand the needs of structural engineers. The author hopes that this book will help bridge that gap.

Important added features of the 2nd edition include greatly expanded discussions of earthwork, fill, compaction control, grading control, inspection, and Special Inspectors. Considerable discussion has also been added regarding the construction and inspection of roadways and parking lots.

The *Workbook* available with the first edition has been discontinued. Test questions are included at the end of each chapter in the 2nd edition, and there is an answer key following the Glossary.

This book will also benefit those who plan to participate in a certification exam on the subject of soils, earthwork, grading, and foundations through gained knowledge on the most important issues relevant to the subjects.

ABOUT THE AUTHOR

Kirby T. Meyer is a professional engineer licensed in Texas and Colorado who has a bachelor's and a master's degree in civil and geotechnical engineering from Texas A&M University and has studied at Harvard and the University of Texas, concentrating in geotechnical and foundation engineering. He has over 40 years' experience in the subjects of this book and is a Fellow of the American Society of Civil Engineers and a Diplomate of the Academy of Geo-Professionals. Contact him with questions and comments at MLAW Consultants and Engineers in Austin, Texas. E-mail address is ktmeyer@mlaw-eng.com.

ACKNOWLEDGEMENTS

Special thanks are due a number of people who have provided invaluable assistance during the writing of this book. Hamid Naderi, P.E., V.P. Product Development, my contact with the International Code Council, for talking me into this project and offering constant guidance and advice; Jay Conner, President of MLAW Consultants & Engineers, for putting up with all the time taken from paying assignments and providing continual encouragement; Dean Read, manager of MLAW's Forensic Department, for making department resources available and providing many helpful comments; Glenn Hadeler with MLAW for helping with the drainage illustrations; Tim Weston V.P., Rick Bowman, P.E.; and others with MLA Labs for providing photographs, illustrations and research for the site investigation and laboratory features. Thanks to all for your help.

A large thank you is due to Keith Brown with MLAW who transcribed the 2nd Edition rewrite (including many revisions), prepared most of the illustrations, and generally kept the whole project on track. Thanks Keith.

Valuable information regarding earthwork operations and equipment was provided by Jim Ivan, Estimator/Project Manager, and Mark Mackenzie, President, both of Ranger Excavating Company in Austin, Texas; Dale Layne, Regional Sales Manager, HOLT CAT, Austin, Texas; Steve Huedepohl, HOLT CAT multimedia communications manager, San Antonio, Texas; Caterpillar, Inc., locations worldwide; Kelly Zerr, HOLT Agribusiness, San Antonio, Texas; Troy Wilson, Wishek Manufacturing, LLC, Wishek, N.D. Thanks to all.

To my wife Janis, who I have known almost as long as I could spell engineer, for encouragement and listening to my stress filled commentary, thank you.

Table of Contents

Chapter 1
INTRODUCTION

Failure of the Transcona grain elevator

Chapter 1

INTRODUCTION

1.1 Why This Book?

This book was conceived as a way to bridge the gap between the sometimes highly theoretical and technically complex world of geotechnical engineering and foundation design on one hand and the ordinary construction activities of actually building foundations and dealing with soil conditions on the other. The book is directed to building code inspectors and others who are involved in permit review and field operations with regard to constructing foundations. References to the 2012 *International Building Code* (IBC) and the 2012 *International Residential Code* (IRC) are provided where applicable. It is assumed the reader who has the duty to enforce the building code has a familiarly with the Preface and Chapter 1 and 2 of 2012 IBC. The titles of referenced standards quoted in the text from organizations other than the International Code Council may be found in Chapter 35 of the 2012 IBC. The referenced standards are a part of the Code when applicable. See 2012 IBC Section 102.4.

1.2 What the Book Is and What It's Not

The book is intended to be a guide and resource for inspectors and construction personnel with regard to soils, earthwork and foundations. Architects, engineers, and engineering students may also benefit from the book. It has sufficient illustrations to provide a level of understanding of the principles involved, and it has sufficient guidelines regarding when it is important to call on professional help, such as a geotechnical engineer or foundation design engineer.

The book is not intended to be a textbook for engineering students, but it may serve as a broad introduction to the subject. It is not a treatise on engineering principles and theories for the trained geotechnical or foundation design engineer; there are a number of excellent books on these topics already available.

1.3 Geotechnical and Geostructural Engineering

Geotechnical Engineering is the current term for what used to be called "soils engineering" or "soil mechanics." It involves knowledge of applied geology, soil formation and composition, reaction of soils to loadings, and the strength or stability capacity of soils. It is also involved with ways of obtaining geotechnical information, testing samples in the laboratory, and preparation of reports that can be used by others for the design of foundations, retaining walls, slope stabilization, and pavements. Therefore, geotechnical engineering is a combination of rough-and-ready field work, precise lab testing, the experience of knowledgeable professionals and technicians, and sometimes quite detailed theoretical analysis. Practioners of geotechnical engineering are typically experienced engineers with advanced degrees in geotechnical engineering or geological engineering.

Geostructural Engineering is a term that refers to design of any structure that is involved with the earth, such as foundations, retaining walls, pavement structures, and stabilization of slopes. It is a combination of the terms "geotechnical" and "structural," forming the word "geostructural." Another name for it, which covers many of the items listed above, is "foundation designing," usually practiced by structural engineers. The best combination of education and experience for designing geostructural works should include a background in geotechnical engineering as well as structural engineering. Since the properties and forces acting on soil can change, depending on a variety of factors such as environmental change and movements of the structures themselves, an important consideration in many soil-related structures is "soil-structure interaction." For example, in the design of retaining walls, slight movement of the top of the wall will change the earth loadings behind it from an "at-rest" state to an "active" state, with the "active" state exerting lower lateral pressures than the "at-rest" state. In other words, the soil and the structure react together to produce load conditions that may be considerably different than if the structure were totally passive and not subject to slight movements.

1.4 Geo-hazards

Building codes all have a common theme—the avoidance of hazards to people or the constructed environment. Put another way, codes are intended to protect persons from injury or death and to protect structures from economic loss. The reasons for protecting persons are obvious and overriding. Economic loss will occur if a structure fails or becomes non-functional. Economic loss will also occur if money is wasted on over-design and unnecessary construction costs.

With regard to soils and foundations, the purpose of the codes and good engineering is to avoid "geo-hazards," which are conditions of the earth that could cause a safety concern or damage to a structure. In general order of severity potential, geo-hazards are:

Earthquake phenomena

- seismic shaking

- soil liquifaction

- seismic slope failures

- fault surface rupture

Non-earthquake phenomena

- static slope failures

- bearing failures

- settlement

- retaining-structure failures

- hydroconsolidation and hydrocollapse

- expansive soils

- frost heave

Earthquake or seismic hazards are generated by rocks within the earth building up stresses and suddenly slipping, causing movement and shaking of the surrounding areas. The shaking can cause structures to collapse, generate tsunamis, and cause slope failures. Some types of saturated soils can liquefy resulting in surface spreading, settlement, or lateral displacement of structures. Roads and utilities can be sheared apart.

Slopes can fail without earthquakes due to water seepage, ill-advised excavations, or extra loads above the slope. Bearing failures are the result of a footing having more load than the soil strength will bear, leading to a shearing failure and footing movement. A geo-hazard called settlement can occur in a static environment because the added weight of the structure squeezes water out of the soil, causing volume reduction.

Retaining structures are intended to hold back earth at a change in soil elevation. Unless properly selected, designed, and constructed to resist horizontal earth loads, they can tilt or collapse.

Hydroconsolidation is the settlement of fills that were fine when placed, but settled after water entered the fill. This usually has to do with the water softening soil particles in the fill, which then collapse into a tighter-packed mass. Hydrocollapse is a similar geo-hazard occurring in natural deposits.

Expansive soils (swelling or shrinking) can move footings and buildings either up or down as a result of moisture change in clay soils. The moisture change can be due to wet or dry weather cycles or due to bad drainage or plumbing leaks. Expansive soil phenomena rarely endanger life, but because of widespread occurrences, they are near the top in causing economic loss.

Results ranging from aggravation to disaster could be the outcome of the responsible person failing to deal with any of the geo-hazards. His or her understanding of soils and foundation construction is vital to avoiding the damage that the earth and mankind's construction can produce. All the geo-hazards are dealt with in the following chapters.

1.5 Test Questions

(All test questions in this book assume the student has available a 4-function calculator)

MULTIPLE CHOICE

1. Geo-hazards that are related to earthquake phenomena are (select three):
   a. slope failure
   b. expanding earth
   c. liquification
   d. fault surface rupture
   e. panic in animals

2. Non-earthquake related geo-hazard phenomena include the following (select three):
   a. bearing failure
   b. expansive soils
   c. ocean tide changes
   d. settlement
   e. air pollution

3. The book is not intended to be a theoretical text book on soils and foundations for use by (select one):
   a. trained geotechnical or foundation design engineers
   b. architects
   c. construction personnel
   d. construction inspectors

4. Geotechnical engineering refers to a discipline that involves (select three):
   a. electrical circuits
   b. knowledge of applied geology
   c. soil formation and composition
   d. reaction of soils to loadings and strength or stability capacity of soils
   e. storm-water drainage

5. Examples of use of geostructural engineering (select two):
   a. analyzing a structure based on geographical maps
   b. design of retaining walls
   c. design of laterally loaded piles
   d. design of beams
   e. design of sanitary sewers

6. In a retaining wall the earth loadings "at rest" are (select one):
   a. lower than the earth loadings in an "active state"
   b. higher than the earth loading in an "active state"
   c. related to wind-induced vibrations

7. Expansive soils can move footings in buildings (select one):
    a. only down
    b. either up or down as a result of moisture change in clay soils
    c. footings cannot be moved

8. Expansive soil phenomena (select two):
    a. rarely endanger life
    b. rate near the bottom in economic loss
    c. rate near the top in economic loss
    d. are not geo-hazards

9. Referenced standards quoted in the code from organizations other than the International Code Council may be found in (select one):
    a. Chapter 16 of 2012 IBC
    b. Chapter 35 of 2012 IBC
    c. 2012 IRC

10. A geo-hazard is a condition of the earth which could (select two):
    a. involve wind damage
    b. cause a safety concern
    c. damage a structure
    d. involve a lightning strike

11. The function of the building codes is (select one):
    a. to maintain traffic control
    b. to ensure health, safety and welfare of the human population
    c. to provide for neat and orderly construction in cities

Chapter 2

THE PURPOSE OF FOUNDATIONS AND A FOUNDATION'S RELATIONSHIP TO SOIL

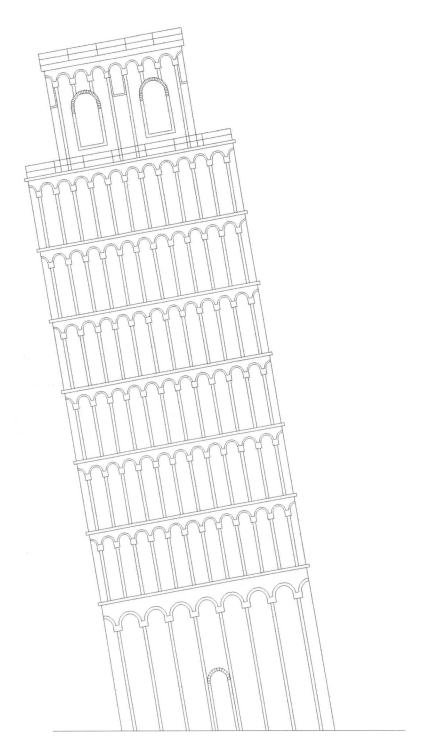

Leaning Tower of Pisa.
The tower has been stabilized 16 feet out of plumb.

Chapter 2

THE PURPOSE OF FOUNDATIONS AND A FOUNDATION'S RELATIONSHIP TO SOIL

2.1 Purpose of Foundations

A foundation's major function is to support the structural loads of the building above and distribute them to the soil or rock of the earth without failure of the soil. Preventing this type of failure is the main reason for determining a safe or allowable bearing pressure, which is the pressure exerted by the foundation per square foot on the supporting soil or rock. Bearing pressure is usually expressed as pounds per square foot (psf) or kips per square foot (ksf): ("kip" is shorthand for kilo + pound, or 1000 pounds.)

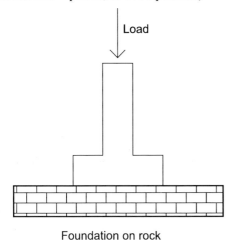

Foundation on rock

Same foundation load on soil

FIGURE 2.1
FOUNDATION ON ROCK AND THE SAME LOAD ON SOIL

2.1.1 Major failure. If the contact or bearing pressures are too great for the capacity of the supporting soil, a shear failure may result. This is an actual tearing or disruption of the supporting soil under the foundation loads, causing major movement of the foundation up to the point of total collapse. See Figure 2.2. A footing might not fail by plunging straight down into the earth, but typically it will tilt in its failure mode. The allowable or "safe" bearing values that can be applied to the soil or rock material supporting the footing are generally determined with a factor of safety of at least three.

2.1.2 Excessive distortion. Another reason for determining the allowable bearing pressure is the prevention of movement of the footings but without a shear failure or total tearing of the soil or rock material. A common phenomenon that will produce this movement is settlement due to consolidation of the soil. During settlement, parts of the foundation can move further downward than other parts, causing distortion of the structure. Settlement can occur in soft soils that have high water contents in which the water is gradually squeezed out of the soil due to applied foundation pressures. The reduction in volume lets the footing move downward. Another cause may be improperly compacted fill underneath the footing. Excessive settlement can cause racking and cracking or tilting of the building. In extreme cases, collapse may occur.

Another type of building movement not involving a bearing capacity failure or settlement is caused by swelling of expansive clay soil. This is a more common problem with lightly-loaded foundations, such as residential foundations, light commercial structures, or lightly-loaded warehouse floors. This is a frequent type of foundation problem in many parts of the Western and Southwestern United States. These areas may have dry expansive clay along with a climatic condition that permits periodic infusions of moisture under the foundation. The result on the structure is the same as bearing failure or settlement because different parts of the foundation move relative to other parts. Sometimes a condition can exist in which settlement and upward expansion both are present, such as in the case of a cut-fill site, where intact swelling clay soil exists on the cut half of the building, and poorly compacted fill may exist on the fill side; see Figures 2.3 and 2.4.

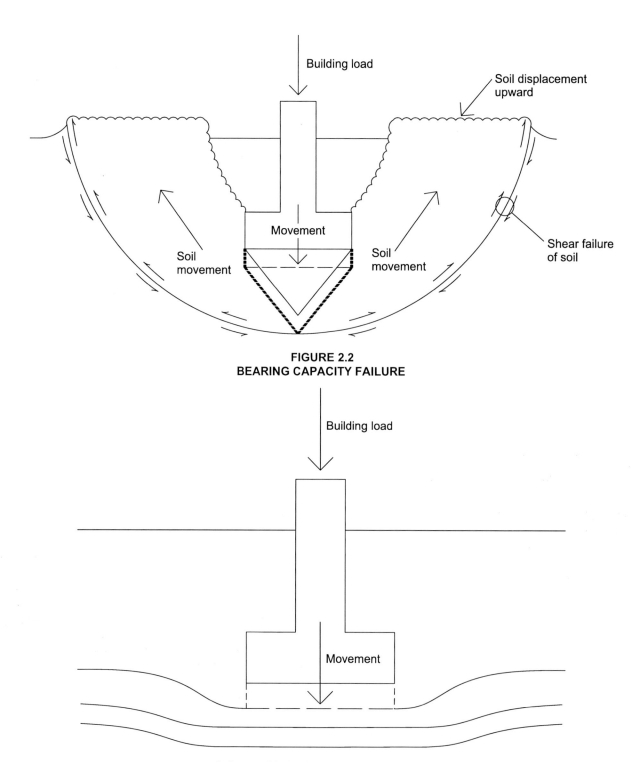

FIGURE 2.2
BEARING CAPACITY FAILURE

Building load

Movement

Soil consolidation by squeezing out water or
squeezing out air due to poorly compacted fill.

FIGURE 2.3
SETTLEMENT WITHOUT SHEAR FAILURE OF SOIL

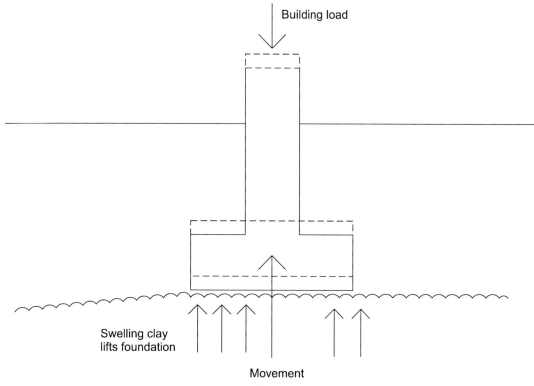

FIGURE 2.4
EXPANSIVE CLAY EFFECTS

Excessive distortion of buildings can also occur because of lateral movement of a sloping site called "downhill creep." This may not be as dramatic as a landslip, which of course would also produce foundation damage, but creep can occur slowly over a long time and eventually cause tilting or distortion of the foundations. This could be a problem for even deep foundations such as piers or piles on sloping sites.

2.1.3 Often the foundation is also the floor. The first floor of a building is often an integral part of the foundation. This is more prevalent in soil-supported slab foundations than other types. In this case the function of a foundation floor is to provide a habitable surface and protect the interior of the building from moisture as well as termites, ants, or other insects that might attempt to gain entrance to the structure.

Moisture is frequently found in soil below foundations, and the entrance of such moisture to the interior of the building should be prevented to avoid damaging floor finishes or producing high humidity, possibly leading to mold infestation or generating damp rot of sill plates. Interior moisture combined with a wood structure also invites termites or other wood-destroying insects.

The foundation should also keep weather-related elements out, including surface and subsurface water in the case of below-grade foundations, such as in basements or split-level construction. Moisture and drainage protection are important considerations for foundations.

2.2 Relationships Among Soil, Rock, and Foundations

It is important to study the soil or rock at a site in relation to foundation design and construction because of the highly differing bearing values possible. The soil can be very soft and be subject to major shearing failures or settlement due to consolidation of soil. Stiff to hard clay soil with good bearing capacity can be the cause of swelling, resulting in structurally damaging heaving. A variety of allowable bearing capacities can be utilized for soil and rock ranging from only a few hundred pounds per square foot up to thousands of pounds per square foot, depending on the soil or rock properties and the configuration of the foundation.

Naturally, any kind of rock is typically stronger than any type of soil, and bearing on rock can easily be secure from shearing failure or settlement, assuming proper knowledge of the strength of the rock is available. Rocks can have a wide variety of safe bearing capacities ranging from 5,000 to over 80,000 psf.

The size and arrangement of foundation elements are greatly dependent on the properties of the soil or rock found at the site as well as the needs of the building structure. Often rock can be found at shallow depths below several feet of soil, and for heavy structures it would be desirable to reach this rock with the foundation elements. Even deep rock may be usable for better bearing and more stable conditions by the use of drilled piers or driven piling. Proper design and construction of foundations cannot be rationally performed without some knowledge of the profile and properties of the soil or rock that are to support the foundation.

2.3 Foundation Loads

To size any foundation element it is necessary to note two things, the capacity of the soil and the load to be applied to the foundation element. Loads on the foundation are loads from the structure. The structure loads are distributed by the designer to contact the earth at various points and by various methods. These contact points with the earth are the foundations.

In a simple case of sizing a footing, the loads coming down onto the footing are divided by the safe bearing pressures of the soil to determine the size of the footing, usually in square feet. From this, the dimensions of the footing can be determined. The same principle holds for all types of foundations, although some may be piles or piers, and some of the footing loads are taken out in soil side skin friction as well as bearing on the bottom of the piers.

The determination of structural loads consists of obtaining the dead load of the structure and adding live loads, roof live loads, rain loads, wind loads, snow loads, and seismic loads. Other load considerations may occur within the structure, for example those loads having to do with the thermal changes of members. These do not usually impact the foundation design. Other loads that may impact the foundation design include horizontal earth loads and hydrostatic loads from ground water or from flooding. A brief description of each of these loads follows, and the reader is referred to Chapter 16 of the 2012 IBC for more detailed treatment.

- **Dead loads** – These are the weights of the materials that make up the building including the walls, floors, roofs, built-in partitions, and the weights of fixed equipment and built-in cabinetry. In other words, these are the loads that are built into the structure and are not readily moved. To determine the dead loading of parts of the structure, the plans should be consulted for sizes and shapes, and a reference made to a source for the unit weight of the various materials. Construction materials range from 490 pounds per cubic foot (pcf) for steel to 150 pcf for concrete and less than 100 pcf for wood materials. The actual weights of the materials to be built or installed and fixed into the structure must be determined and added together.

- **Live loads** – Live loads are those produced by the use and occupancy of the building, such as people loads and movable furniture. These are typically determined by tables in the building code, such as Table 1607.1 in the 2012 IBC. Live loads are based on a uniform distribution in psf over the floor or roof areas; in some cases a concentrated load is given for certain usages. The distributed uniform load and the concentrated load are analyzed to see which will produce the greatest effect on the structure, and that number is used.

- **Live loads (roof)** – These are loads produced during maintenance by workers, equipment, and materials on the roofs and possibly by people or planters on the roof; a garden-type roof area is an example.

- **Rain loads** – Rain loads are similar to snow loads in that they are the result of weather-related phenomena. Rain is generally a problem on roofs that are flat or nearly flat (this type of roof construction is generally a bad idea anyway). Flat roofs have drain systems that can be clogged with leaves or debris and should have a secondary drainage system in case the primary system is clogged. The depth of the primary system plus the secondary system overflow depth is calculated by an equation in Section 1611.1 of the 2012 IBC. A dangerous condition for flat roofs is ponding wherein the roofs have insufficient rigidity and deflect downward in mid-spans as the weight of water increases. A number of notable roof failures have occurred due to water ponding on flat roofs that were deflecting, creating a bowl effect. This bowl effect is increased as more rain falls and the roof deflects further, ponding more water and generating a vicious cycle that can and sometimes does lead to collapse. Section 1611.2 of the 2012 IBC advises that ponding instability shall be evaluated in accordance with 8.4 of ASCE 7.

- **Snow loads** – Snow loads are those resulting from accumulation of snow on a roof. Snow loads are based on Figure 1608.2 of the 2012 IBC and Chapter 7 of ASCE 7. The base load may be modified depending on the ability of a roof to build up drifts greater than the base depths shown on the charts. Warmer climates naturally have lower snow loadings than colder climates, in general.

- **Wind loads** – The loading on structures and the foundations is influenced by horizontal, uplift, and downforce loads from the wind. Due to the overturning effect of wind, the taller the structure, the greater is the foundation loading on the downstream edge and uplift forces on the upstream side of the foundations. According to Section 1609.1.1 of the 2012 IBC, wind loads are to be determined in accordance with Chapters 26 through 30 of ASCE 7 or Section 1609.6 of the IBC. The faster the wind blows, the greater the horizontal force, but in addition there are shape factors and exposure categories depending on the structure's site and height above ground. Wind loading for structural design and safety of the superstructure is an important consideration. However, the horizontal, downforce, and uplift forces applied to the foundations are of primary interest for the foundation designer.

- **Seismic loads** – These are loadings due to earthquakes shaking the ground on which the structure is established. They are summarized in Section 1603.1.5 and discussed in detail in Section 1613 of the 2012 IBC. In this book, Chapter 10 deals with seismic conditions. The determination of seismic loadings, both vertical and horizontal, on foundations is complex, and a detailed discussion of this is beyond the scope of this book.

- **Flood and hydrostatic loads** – Flood loads are generally hydrostatic water pressures based on the unbalanced height of the water that is applied to a structure. They can also include current velocities that would increase the horizontal loading. Hydrostatic loads may not be in a floodplain but could simply be ground water that has built up against a retaining wall, or they could be loads of fluid within the structure. If the structure was an industrial facility and the fluid levels were variable, this could be a form of a variable live load. Flood loads are discussed in Section 1612 of the 2012 IBC.

- **Soil lateral loads** –Section 1610 of the 2012 IBC discusses lateral loads on basements, foundations, and retaining walls. These loads are discussed in this book in Chapter 18.

Table 1607.1 of the 2012 IBC lists the code-mandated live loads for various types of construction. See Table 2.1 on the next page.

Section 1605 of the 2012 IBC deals with load combinations, and two types of design loading procedures are mentioned. One procedure is the strength design or load and resistance factor design, sometimes abbreviated as LRFD. The LRFD applies a factor of safety to the loadings and analyzes all the materials and elements of construction at their ultimate load capacity. The other procedure is the allowable stress design loading using working stress and service loads. In this procedure structural elements are designed using the actual loads expected in service, and a factor of safety is applied to the materials, such as concrete, steel, or wood.

Common practice in the design of foundations using soil-bearing capacities is to use the second procedure or the allowable stress design process, which is discussed in more detail in Section 1605.3 of the 2012 IBC. It is important for the foundation designer to know what the recommendations in the geotechnical report are based on, whether they are ultimate strength capacities of the soil materials under various types of loadings or whether they are allowable pressures that have a safety factor built in. This can be confusing and must be absolutely clear before design can proceed.

The various loading procedures described above bring into focus a difference in mind-set between structural designers and geotechnical engineers. Structural design is quite complex; however, the designer can specify the materials, including the strength and other properties, and expect to get them on the job with a reasonable degree of certainty. Then, with the code-mandated live loads, calculated dead loads, wind loads, and other loads, the analysis of the superstructure is a matter of mechanics following sometimes-complex formulas. On the other hand, the soils engineer or geotechnical engineer must take nature as it is presented, and he or she has very little control over the actual strength of the soil, but must simply work diligently to try to determine what is in place and what the soil's strength will be under service loadings. This generates a little more fuzzy approach in geotechnical engineering than is generally expected in structural engineering, and the procedure relies on the judgment of the geotechnical analyst to a greater degree. Geotechnical engineering practitioners spend a major part of their time trying to determine what the properties of the soil are, whereas the structural designer simply specifies what strength of material she or he wants and goes on with the design.

TABLE 2.1
**MINIMUM UNIFORMLY DISTRIBUTED LIVE LOADS, L_o, AND
MINIMUM CONCENTRATED LIVE LOADS[g]**
(2012 IBC Table 1607.1)

OCCUPANCY OR USE	UNIFORM (psf)	CONCENTRATED (lbs.)
1. Apartments (see residential)	—	—
2. Access floor systems		
Office use	50	2,000
Computer use	100	2,000
3. Armories and drill rooms	150[m]	—
4. Assembly areas		
Fixed seats (fastened to floor)	60[m]	
Follow spot, projections and control rooms	50	
Lobbies	100[m]	—
Movable seats	100[m]	
Stage floors	150[m]	
Platforms (assembly)	100[m]	
Other assembly areas	100[m]	
5. Balconies and decks[h]	Same as occupancy served	—
6. Catwalks	40	300
7. Cornices	60	—
8. Corridors		
First floor	100	
Other floors	Same as occupancy served except as indicated	—
9. Dining rooms and restaurants	100[m]	—
10. Dwellings (see residential)	—	—
11. Elevator machine room grating (on area of 2 inches by 2 inches)	—	300
12. Finish light floor plate construction (on area of 1 inch by 1 inch)	—	200
13. Fire escapes	100	
On single-family dwellings only	40	—
14. Garages (passenger vehicles only)	40[m]	Note a
Trucks and buses	See Section 1607.7	
15. Handrails, guards and grab bars	See Section 1607.8	
16. Helipads	See Section 1607.6	
17. Hospitals		
Corridors above first floor	80	1,000
Operating rooms, laboratories	60	1,000
Patient rooms	40	1,000
18. Hotels (see residential)	—	—
19. Libraries		
Corridors above first floor	80	1,000
Reading rooms	60	1,000
Stack rooms	150[h, m]	1,000
20. Manufacturing		
Heavy	250[m]	3,000
Light	125[m]	2,000
21. Marquees	75	—
22. Office buildings		
Corridors above first floor	80	2,000
File and computer rooms shall be designed for heavier loads based on anticipated occupancy	—	—
Lobbies and first-floor corridors	100	2,000
Offices	50	2,000

(continued)

TABLE 2.1—continued
**MINIMUM UNIFORMLY DISTRIBUTED LIVE LOADS, L_o, AND
MINIMUM CONCENTRATED LIVE LOADS[g]**
(2012 IBC Table 1607.1)

OCCUPANCY OR USE	UNIFORM (psf)	CONCENTRATED (lbs.)
23. Penal institutions		
Cell blocks	40	—
Corridors	100	
24. Recreational uses:		
Bowling alleys, poolrooms and similar uses	75[m]	
Dance halls and ballrooms	100[m]	
Gymnasiums	100[m]	—
Reviewing stands, grandstands and bleachers	100[c, m]	
Stadiums and arenas with fixed seats (fastened to floor)	60[c, m]	
25. Residential		
One- and two-family dwellings		
Uninhabitable attics without storage[i]	10	
Uninhabitable attics with storage[i, j, k]	20	
Habitable attics and sleeping areas[k]	30	—
All other areas	40	
Hotels and multifamily dwellings		
Private rooms and corridors serving them	40	
Public rooms[m] and corridors serving them	100	
26. Roofs		
All roof surfaces subject to maintenance workers		300
Awnings and canopies:		
Fabric construction supported by a skeleton structure	5 nonreducible	
All other construction	20	
Ordinary flat, pitched, and curved roofs (that are not occupiable)	20	
Where primary roof members are exposed to a work floor, at single panel point of lower chord of roof trusses or any point along primary structural members supporting roofs:		
Over manufacturing, storage warehouses, and repair garages		2,000
All other primary roof members		300
Occupiable roofs:		
Roof gardens	100	
Assembly areas	100[m]	
All other similar areas	Note 1	Note 1
27. Schools		
Classrooms	40	1,000
Corridors above first floor	80	1,000
First-floor corridors	100	1,000
28. Scuttles, skylight ribs and accessible ceilings	—	200
29. Sidewalks, vehicular drive ways and yards, subject to trucking	250[d, m]	8,000[e]

(continued)

TABLE 2.1—continued
MINIMUM UNIFORMLY DISTRIBUTED LIVE LOADS, L_o, AND MINIMUM CONCENTRATED LIVE LOADS[g]
(2012 IBC Table 1607.1)

OCCUPANCY OR USE	UNIFORM (psf)	CONCENTRATED (lbs.)
30. Stairs and exits One- and two-family dwellings All other	 40 100	 300 [f] 300 [f]
31. Storage warehouses (shall be designed for heavier loads if required for anticipated storage) Heavy Light	 250[m] 125[m]	 —
32. Stores Retail First floor Upper floors Wholesale, all floors	 100 75 125[m]	 1,000 1,000 1,000
33. Vehicle barriers	See Section 1607.8.3	
34. Walkways and elevated platforms (other than exitways)	60	—
35. Yards and terraces, pedestrians	100[m]	—

For SI: 1 inch = 25.4 mm, 1 square inch = 645.16 mm^2,
1 square foot = 0.0929 m^2,
1 pound per square foot = 0.0479 kN/m^2, 1 pound = 0.004448 kN,
1 pound per cubic foot = 16 kg/m^3.

a. Floors in garages or portions of buildings used for the storage of motor vehicles shall be designed for the uniformly distributed live loads of Table 1607.1 or the following concentrated loads: (1) for garages restricted to passenger vehicles accommodating not more than nine passengers, 3,000 pounds acting on an area of 4.5 inches by 4.5 inches; (2) for mechanical parking structures without slab or deck that are used for storing passenger vehicles only, 2,250 pounds per wheel.

b. The loading applies to stack room floors that support nonmobile, double-faced library book stacks, subject to the following limitations:
   1. The nominal bookstack unit height shall not exceed 90 inches;
   2. The nominal shelf depth shall not exceed 12 inches for each face; and
   3. Parallel rows of double-faced book stacks shall be separated by aisles not less than 36 inches wide.

c. Design in accordance with ICC 300.

d. Other uniform loads in accordance with an approved method containing provisions for truck loadings shall also be considered where appropriate.

e. The concentrated wheel load shall be applied on an area of 4.5 inches by 4.5 inches.

f. The minimum concentrated load on stair treads shall be applied on an area of 2 inches by 2 inches. This load need not be assumed to act concurrently with the uniform load.

g. Where snow loads occur that are in excess of the design conditions, the structure shall be designed to support the loads due to the increased loads caused by drift buildup or a greater snow design determined by the building official (see Section 1608).

h. See Section 1604.8.3 for decks attached to exterior walls.

i. Uninhabitable attics without storage are those where the maximum clear height between the joists and rafters is less than 42 inches, or where there are not two or more adjacent trusses with web configurations capable of accommodating an assumed rectangle 42 inches in height by 24 inches in width, or greater, within the plane of the trusses. This live load need not be assumed to act concurrently with any other live load requirements.

(continued)

j. Uninhabitable attics with storage are those where the maximum clear height between the joists and rafters is 42 inches or greater, or where there are two or more adjacent trusses with web configurations capable of accommodating an assumed rectangle 42 inches in height by 24 inches in width, or greater, within the plane of the trusses.
   The live load need only be applied to those portions of the joists or truss bottom chords where both of the following conditions are met:
   i. The attic area is accessible from an opening not less than 20 inches in width by 30 inches in length that is located where the clear height in the attic is a minimum of 30 inches; and
   ii. The slopes of the joists or truss bottom chords are no greater than two units vertical in 12 units horizontal.

The remaining portions of the joists or truss bottom chords shall be designed for a uniformly distributed concurrent live load of not less than 10 lb./ft^2.

k. Attic spaces served by stairways other than the pull-down type shall be designed to support the minimum live load specified for habitable attics and sleeping rooms.

l. Areas of occupiable roofs, other than roof gardens and assembly areas, shall be designed for appropriate loads as approved by the building official. Unoccupied landscaped areas of roofs shall be designed in accordance with Section 1607.12.3.

m. Live load reduction is not permitted unless specific exceptions of Section 1607.10 apply.

2.4 TEST QUESTIONS

MULTIPLE CHOICE

1. On a cut-fill site in expansive clays, the following can occur (select two):
   a. no changes due to balancing
   b. swelling
   c. dying trees
   d. settlement
   e. better vegetation

2. To size any foundation element it is necessary to note two things (select two):
   a. loading
   b. inspection
   c. reinforcing steel grade
   d. soil capacity

3. If the load applied to the footing is 20 kips and the allowable bearing capacity is 1500 psf, the required area of the footing is (select one):
   a. 10.6 sq. ft.
   b. 13.3 sq. ft.
   c. 1.33 sq. ft.
   d. 100 sq. ft.

4. A foundation's major function is to (select one):
   a. re-direct surface water
   b. support the structural loads from the building above and distribute them to the soil or rock of the earth
   c. make the lower portion of the building heavier

5. Soft soils that have high water contents (select one):
    a. have high bearing capacity
    b. are subject to the water being squeezed out of the soil resulting in settlement
    c. are highly unlikely to settle
    d. may penetrate through basement floors

6. Downhill creep (select one):
    a. can produce results more dramatic than a land-slip
    b. is not a problem for deep foundations such as piles
    c. may cause long term damage to structures
    d. can be stopped by planting trees

7. The entrance of moisture from the soil below foundations into the interior of the building (select two):
    a. is never a problem
    b. can cause damp rot to wall sill plates
    c. can damage floor finishes
    d. can soften ceramic tile

8. The allowable bearing capacities that can be utilized for shallow footings in soil and rock range (select one):
    a. from less than 1000 psf up to many thousands of psf
    b. is not important since they are not covered in the codes
    c. can be 50 psf
    d. is usually calculated without considering a factor of safety

9. Rocks can have a wide variety of safe bearing capacities (select one):
    a. ranging to over 80,000 psf
    b. ranging up to 250,000 psf
    c. of less than 1,000 psf

10. Design and construction of foundations, to be rationally performed (select one):
    a. must be based on information found on topographical maps
    b. cannot be performed without some knowledge of the properties of the soil or rock that supports the foundation
    c. can be performed by relying on contractors' opinions

11. Table 1607.1 in the 2012 IBC lists (select one):
    a. dead loads for use in design of buildings
    b. live loads for use in design of buildings
    c. lateral earth pressures minimum concrete strength

Chapter 3
ROCK

Limestone of the Glen Rose Formation, Cretaceous Age, seen in a highway cut near Austin, Texas

Chapter 3
ROCK

3.1 Formations

Formations are geologic units that have been mapped by extensive studies by geologists using field work, aerial photography, boring information, and other sources. Formations are generally identified by their geologic age and usually have a fairly consistent set of attributes in a local area. For this reason it is quite useful for the engineer to understand the geologic formations in a local area, since test-boring data can be extended between borings if the engineer knows the geology and the properties of the formations. Most formations consist of rock of various sorts. Geologists tend to call every formation a rock, even if it is soft clay shale, unless it is a surface deposit that has been weathered into something that can grow vegetation.

Some formations consist of very hard rocks that are normally quite good support for foundations and other earth-related structures. However, they can be very difficult and costly to excavate. Rock can include igneous deposits such as granite or sedimentary deposits such as limestone or shale. Igneous rocks are those that are formed from molten rock (lava or magma). Sedimentary rocks are formed when mineral precipitates, or shells of marine organisms are deposited in water and are hardened by pressure or cementation over long time periods. Metamorphic rock is another major rock type. These rocks are formed from any of the igneous or sedimentary rocks by transformation from heat and pressure. Of course within each of these types of rocks there are quite large variations in strength and excavation difficulty. Some of the more difficult rock formations to evaluate for foundation construction purposes can be those of layers of limestone or sandstone intermingled with layers of clay or clay-shale.

A useful engineering definition of rock is a material that will not "slake" when partially submerged in water. This indicates a virtually irreversible cementation, which will allow the rock to retain its structure throughout its use. Another way of determining rock is if it has a brittle fracture or if a small flat piece can be broken with a detectible "snap." The definition of shale is especially confusing at times. Shale is generally considered to be a material meeting the rock tests named above but also is thinly layered or "fissile." Shale can weather into clay, and frequently there is a change in the profile with depth, with weathered clay-like materials near the surface and the more rock-like shale at deeper depths.

3.2 Geology

Geology is the science of the origin, formation, and description of materials of the earth. Geologists have separated various formations and rock units into time periods ranging back to nearly the origin of the earth up to the present time. Figure 3.1 is a representative time scale of geologic units.

FIGURE 3.1
GENERALIZED GEOLOGICAL SEQUENCE
(oldest at bottom)

GROUP	SYSTEM		APPROXIMATE AGE (YEARS)
Recent	Quaternary	Present Day	Present – 10,000
		Pleistocene	10,000 – 1.6 Million
Cenozoic	Tertiary	Pliocene Miocene Oligocene Eocene	1.6 – 65 Million
Mesozoic	Cretaceous Jurassic Triassic		65 – 250 Million
Paleozoic	Permian Carboniferous Devonian Silurian Ordovician Cambrian		250 – 570 Million
Precambrian			Origins of Earth to Cambrian

Any of the rock units shown on the time scale could be exposed near the ground surface at any particular locality depending on the erosional or faulting history of the area. Frequently, regardless of what ancient formation exists at a fairly shallow depth, there are recent deposits overlying the material. Recent deposits could be as old as 1.6 million years or as young as yesterday. These are typically deposits due to water action. Deposition of soil derived from erosion sources often includes river or creek alluvium (alluvium means water-deposited), side bank deposits or deltas, such as the Mississippi River Delta, or wide areas of waterborne deposits called terrace or sheet deposits. Sometimes colluvium, which is material that has fallen from a slope or bluff by gravity, may be found covering older deposits.

Geotechnical engineering is mainly concerned with the engineering properties (strength, compressibility, shrink-swell, permeability) of the upper layers of the Earth's crust ranging from a few feet to perhaps several hundred feet, depending on the type of deposit and the nature of the proposed foundation.

3.3 Faults

Faults are fractures in rock formations caused by relative movements of the Earth's crust. Some faults are famous, such as the San Andreas Fault in California, because their movements result in frequent earthquakes. An earthquake occurs when stresses build up in rock, and the stresses are suddenly relieved by slippage, creating a large acceleration and movement of the surrounding area. The stress build-up results from large plate movements of the Earth's crust, deep molten rock movements, or area subsidence. Faults can be of various types. They can be "strike-slip" movement from side to side or "thrusting" faults in which a sloping layer is forced under or over an adjacent layer. Active faults that tend to move with a sudden acceleration from time to time are of great interest to foundation and structural designers since they produce earthquakes, which can destroy or severely damage structures. If earthquakes occur within the ocean, tsunamis may result, which can be more damaging than earthquakes and result in greater loss of life.

In many areas there are numerous faults that are not active, meaning that they have not moved or caused earthquake-type accelerations for very long periods of time. Sometimes, for foundation design, even these "dormant" faults can cause problems because there may be a rock formation outcropping near the surface immediately adjacent to a clay formation. This will cause two different types of materials to be next to each other at the ground surface, which could cause the reaction of a foundation to be very different in different parts.

Faults can be clean breaks, which are very narrow, or they can have a wide zone of ground-up and mixed-up material. Often faults are sources of springs, and many major springs around the world are located at a fault zone because aquifers have been brought to the surface and fractured open. Figure 3.2 illustrates various kinds of faults.

3.4 Rock as a Foundation Support

Rock is an excellent material for supporting foundations because of the high strength and consequent high bearing values that can be assigned to this material. Naturally, the higher the bearing value of a material, the smaller a foundation can be to distribute the loads of the structure to the earth. Rock strength comes in many grades, ranging from rock that is only somewhat stronger than a hard clay up to material that may be stronger than concrete. To apply standardized permissible bearing values to rock, the rock materials must be classified carefully. Table 3.1, showing presumptive bearing values for rock and soil, is taken from the 2012 IBC. The 2012 IRC has a similar listing in Table R401.4.1.

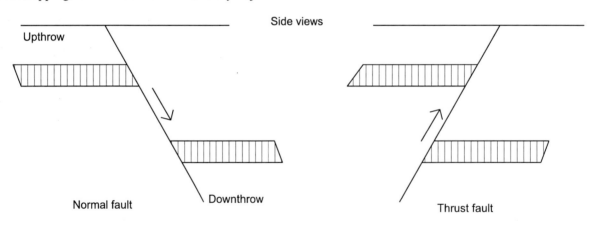

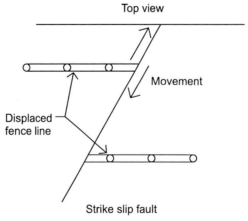

FIGURE 3.2
TYPES OF FAULTS

**TABLE 3.1
ALLOWABLE FOUNDATION AND LATERAL PRESSURE
(TABLE 1806.2, 2012 IBC)**

CLASS MATERIALS	ALLOWABLE FOUNDATION PRESSURE (PSF)[D]	LATERAL BEARING (PSF/F BELOW NATURAL GRADE)[D]	LATERAL SLIDING	
			COEFFICIENT OF FRICTION[A]	COHESION (PSF)[B]
1. Crystalline bedrock	12,000	1,200	0.70	---
2. Sedimentary and foliated rock	4,000	400	0.35	---
3. Sandy gravel and/or gravel (GW and GP)	3,000	200	0.35	---
4. Sand, silty sand, clayey sand, silty gravel and clayey gravel (SW, SP, SM, SC, GM and GC)	2,000	150	0.25	---
5. Clay, sandy clay, silty clay, clayey silt, silt and sandy silt (CL, ML, MH and CH)	1,500	100	---	130

For S1: 1 pound per square foot = 0.0479 kPa, 1 pound per square foot per foot = 0.157 kPa/m.

a. Coefficient to be multiplied by the dead load.

b. Cohesion value to be multiplied by the contact area, as limited by Section 1806.3.2.

c. Where the building official determines that in-place soils with an allowable bearing capacity of less than 1,500 psf are likely to be present at the site, the allowable bearing capacity shall be determined by a soils investigation.

d. An increase of one-third is permitted when using the alternate load combinations in Section 1605.3.2 that include wind or earthquake loads.

It is also important for the designer and the inspector to be certain that a foundation with high bearing pressures is not established on a thin layer of rock, which may have a softer material below it and therefore would be subject to a punching failure. The presumptive bearing values given in Table 3.1 above are conservative numbers, provided the rock or soil classification is reasonably accurate. To determine more precise and possibly higher bearing values for rock formations, a geotechnical investigation is necessary in which cores of the rock are obtained, lab testing is utilized, and some theoretical calculations are done to determine what the rock will safely carry. Higher safety factors are used for rock allowable-bearing pressures than for soil because of the uncertain effect of fractures or joints in rock.

3.5 TEST QUESTIONS

MULTIPLE CHOICE

1. Sedimentary deposits are those that are formed from (select one):
    a. molten rock (lava or magma)
    b. water deposits
    c. fill placed by trucks

2. Geology is the science of (select one):
    a. mapping streets and towns in a locality
    b. finding the best highway route
    c. the origin, formation, and description of materials of the Earth
    d. evolution

3. Select the correct statement (select one):
    a. The Cretaceous Age is older than the Jurassic Age
    b. In normal geologic sequence the oldest formations are at the top of the profile
    c. Recent deposits can be as old as 1.6 million years
    d. Thrusting faults move from side to side

4. Sudden movements of faults (select two):
    a. may result in earthquakes
    a. cause chickens to stop laying eggs
    a. may cause tsunamis
    a. are the result of tornados

5. Faults which have not been active for very long periods of time (select one):
    a. never pose problems with regard to foundation design
    b. may spontaneously collapse
    c. could cause two different soil materials to be present at the ground surface

6. The presumptive allowable foundation pressures shown in in the building code (select one):
    a. can never be exceeded even with the use of a geotechnical study of the site
    b. are safe to use if the bearing material is identified correctly
    c. may not be safe to use

7. Formations are generally identified by (select one):
    a. age
    b. color
    c. the presence of springs

8. Alluvium means soils deposited (select one):
   a. by volcanic ash
   b. from erosion
   c. from airborne dust

9. Higher safety factors are used for rock allowable bearing pressures than for soil because of the uncertain effect of (select one):
   a. core holes
   b. acids generated by tree roots
   c. joints and fractures
   a. concrete interaction with rock

10. Table 1806.2 of the IBC permits allowable foundation pressures of (select one):
    a. 4,000 psf for crystalline bedrock
    b. 400 psf for sedimentary rock
    c. 3,000 psf for sandy gravel or gravel

Chapter 4
SOIL

A clay soil that has a high capability to shrink or swell when moisture content is changed

Chapter 4
SOIL

4.1 Origin

In general, soil originates from a rock of some sort. Soil typically forms on the surface of rocks by natural weathering processes, which include physical breakup due to freezing and thawing, or drying (shrinking) and wetting (swelling) of rocks near the surface. The large pieces are then further broken down by the same processes, and oxidation can take place from contact with the oxygen in air or water. Weak acids may form in rain, which will decompose limestone. Once decomposition from all sources is sufficiently advanced, a near-surface soil is produced that is able to support organic life, which hastens the decomposition by further breaking down the weathered particles and supplying more organic material. The speed and depths of such weathering processes depends a great deal on the climate and the original rock type. Limestone and granite rocks will both follow the above process but produce different soils with different mineral compositions. Most soil

will eventually weather out to form clay or silt soils. Figure 4.1 shows two frequently encountered rock and soil profiles.

Soils can be moved around primarily by water but also by wind and glaciers. Soils moved by water or glaciers can be moved off uplands or mountains and the partially weathered material further subjected to grinding and degradation while being moved by stream flows in creeks and rivers. Contact of rock particles with moving glacier ice can also apply major grinding action. Creeks and rivers will sort the material, with the heavier gravel sizes being deposited as soon as the current slows down enough and the finer silt and clay sizes being carried longer distances. This is the main reason that the Mississippi is called the "Big Muddy." Creeks and rivers will deposit the sand and gravel along the sides of the main stream where the current velocity slows down during floods, while silt and clay are carried on in suspension. At some point the velocity of water slows enough, such as in river deltas, to allow the silt and clay to

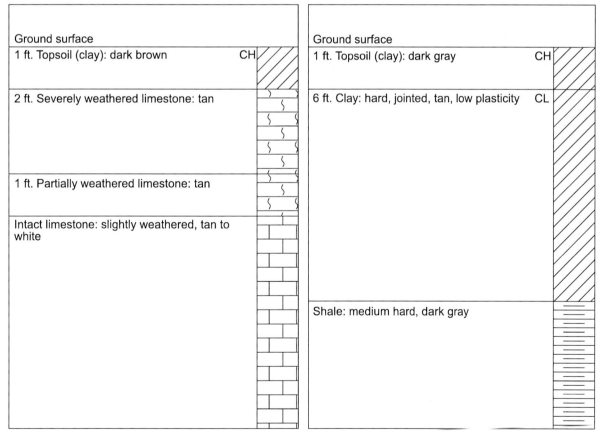

For SI. 1 foot = 304.8 mm.

FIGURE 4.1
TYPICAL SOIL PROFILES OVER ROCK

be deposited. The Mississippi River Delta is thousands of feet thick in the Gulf of Mexico and consists primarily of silt and clay.

Figure 4.2 shows a geologic outcrop map with older formations overlain by more recent terrace and alluvium deposits (Qt and Qal).

Wind can move soil (a dust storm, for example) resulting in deposits called loess or dunes. Loess deposits can be tricky from a geotechnical point of view since they often appear as stable vertical cliffs. They have adequate strength until they are saturated, at which point they can dramatically collapse. Wind can move sand dunes long distances, sometimes covering structures or even entire settlements. Dust storms can cover many thousands of square miles, leaving fine deposits. Sometimes the deposit is quite thick, and the deposited dust is the basis of much of the soil at a site.

Volcanic eruptions can throw dust-size particles into the air, which can be carried hundreds or even thousands of miles before settling to the earth, often forming a significant thickness of soil.

4.2 Classification

For identification and analysis purposes, several soil classification systems have been developed. These classification systems have a definite method of determining the classification, generally through the grain sizes of the soil material and plasticity factors. The most widely used classification system for engineering purposes is the Unified Soil Classification System (USCS). This was originally called the "Airfield Classification System" and is currently used by the IBC and IRC as well as the U.S. Army Corps of Engineers and many other engineering organizations.

This classification system does not depend on the condition of the soil, such as how wet it is or how compact it is, but is a basic classification of the raw material properties. The USCS is systematic and breaks the soil constituents into coarse-grain and fine-grain material. Coarse-grain material is further divided into boulders, cobbles, gravel, and sand. All of the coarse-grain materials are larger than that which will pass the #200 sieve, which has openings of 0.074 millimeters. This is a size that is barely visible to the naked eye, and sizes smaller than this are not visible as individual grains. Soils are called coarse-grained soils if

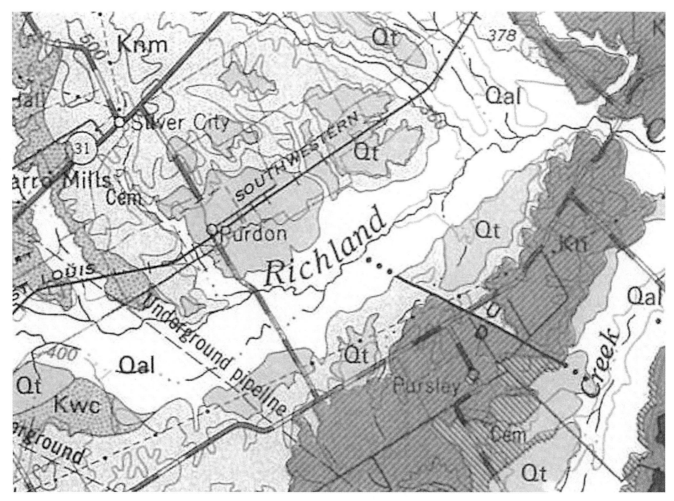

Courtesy of Bureau of Economic Geology, The University of Texas at Austin

FIGURE 4.2
GEOLOGIC OUTCROP MAP WITH RECENT DEPOSITS (Qt AND Qal) OVER CRETACEOUS AGE ROCKS

more than half is retained on the #200 sieve. Of course, mixtures of fine-grain and coarse-grain soil frequently exist in nature, so basic coarse-grain soils classifications often have a fine-grain description added.

The coarse-grain division is as follows:

- Boulders – Anything larger than twelve inches in diameter

- Cobbles – Material from 3 inches to 12 inches in diameter

- Gravel – 3 inches down to approximately $^1/_4$ inch

- Sand – $^1/_4$ inch or passing the #4 sieve down to the #200 sieve size

Coarse-grain material carries an upper-case G for gravel or an upper-case S for sand in the front of the classification symbol. For example, concrete coarse aggregate is divided into gravel and sand. Gravel, the largest concrete aggregate, is typically not bigger than $1^1/_2$ inches to 2 inches and goes down to $^1/_4$ inch. Typical concrete sand ranges from about $^1/_4$ inch down to a very small amount passing the #200 sieve. The amount of sand passing the #200 sieve is on the order of 0 to 3 percent in a typical concrete mix.

Coarse-grain soils are further classified as to whether or not they are well graded or poorly graded; the classification is indicated by the symbol of W or P. Either of these symbols indicate that the soil is relatively clean and does not have more than 5 percent passing the #200 sieve. A well-graded (W) gravel or sand mixture will have approximately the same percentage of the various sieve sizes available to make a complete distribution. Poorly-graded (P) materials have some sizes missing or have a predominance of only a few sizes. Such materials are also sometimes called "gap graded."

Fine-grain soils are those that pass the #200 sieve, and they can be either silts (M), clays (C), or mixtures. If more than half of a total sample by dry weight passes the #200 sieve, the soil is called fine-grained soil. The further classification of fine-grain soil is determined by combinations of the Atterberg Limits: the Liquid Limit, Plastic Limit, and Plasticity Index. See 7.3.1.3 for a discussion of the Atterberg Limits. These determinations are made by intersecting the Liquid Limit and the Plasticity Index plotted on a "Plasticity Chart" as either a high plasticity clay (CH), a high plasticity silt (MH), a low plasticity clay (CL), or a low plasticity silt (ML). Figure 4.3 is the Plasticity Chart. The entire procedure for soil classification by the USCS is shown below in Figure 4.4.

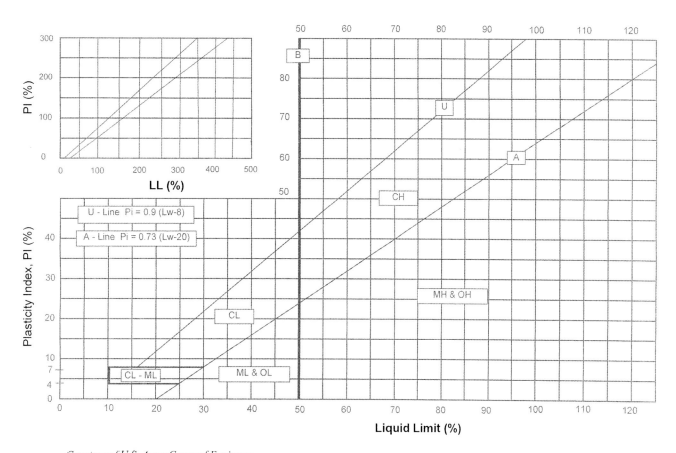

Courtesy of U.S. Army Corps of Engineers

**FIGURE 4.3
PLASTICITY CHART**

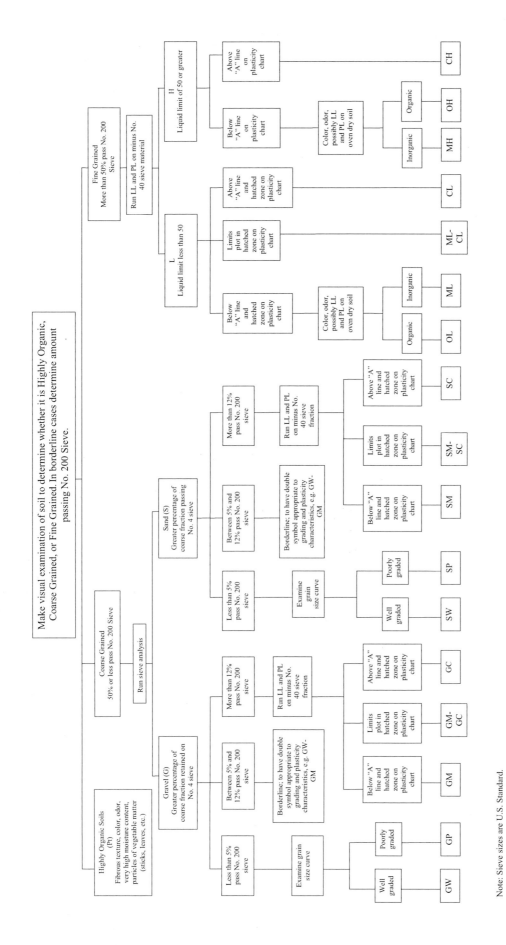

FIGURE 4.4
UNIFIED SOIL CLASSIFICATION PROCEDURE

Courtesy of U.S. Army Corps of Engineers

Note: Sieve sizes are U.S. Standard.

* If fines interfere with free-draining properties, use double symbol such as GW-GM, etc.

There are several other classification systems in use, including agriculture systems and the American Association of State Highway and Transportation Officials (AASHTO) systems. However, for most purposes the Unified Soil Classification System is satisfactory and is in the widest use.

One additional classification is added to the USCS called "organics." An organic soil has a large amount of organic (plant) material and can be especially troublesome for foundation support because of the high water-holding capacity of this material, leading to high potential for settlement. Organic soils carry the symbol "O" such as OH or OL. An extreme example of organic materials would be peat deposits (symbol Pt), which are almost all plant materials. High organic materials are generally unsuitable for foundation use.

See ASTM Standard D 2487 – 06 (refer to Chapter 35 of the 2012 IBC) for a more detailed treatment of the USCS classification procedures. Section 1803.5.1 of the 2012 IBC requires the use of the ASTM standard. Table R405.1 of the 2012 IRC lists engineering properties related to the Unified Soil Classification System (see Table 4.1).

The soil classifications can be determined precisely by laboratory methods, such as washing the materials through a nest of sieves of decreasing size and determining the dry weight of each retained fraction expressed as a percentage of the total sample. In addition, the fine-grain materials can be tested by the Atterberg Limits. The various laboratory tests are described in 7.3. However, there are ways to use "field-expedient" methods to classify soils. These methods can be reasonably accurate in the hands of an experienced person. Coarse-grain soils that are not too tightly bound with fine-grain material can be spread out on a canvas or a hard surface and the various sizes separated by hand to get a rough idea of the percentages. Fine-grain soils can be classified as having silt or clay characteristics by using simple tests in the field. The field-expedient estimates of soil classification and other properties of soil are described in more detail in Section 7.4.

4.3 Bearing and Strength

Soil bearing value can be determined by presumptive classification tables, such as those found in building codes. Section 1808 of the 2012 IBC, including Table 1806.2, and Section R403 of the 2012 IRC discuss footing design. Alternatively, geotechnical reports analyzing field and lab-

TABLE 4.1
PROPERTIES OF SOILS CLASSIFIED ACCORDING TO THE UNIFIED SOIL CLASSIFICATION SYSTEM
(2012 IRC: TABLE R405.1)

SOIL GROUP	UNIFIED SOIL CLASSIFICATION SYSTEM SYMBOL	SOIL DESCRIPTION	DRAINAGE CHARACTERISTICS a	FROST HEAVE POTENTIAL	VOLUME CHANGE POTENTIAL EXPANSION b
Group I	GW	Well-graded gravels, gravel sand mixtures, little or no fines	Good	Low	Low
	GP	Poorly graded gravels or gravel sand mixtures, little or no fines	Good	Low	Low
	SW	Well-graded sands, gravelly sands, little or no fines	Good	Low	Low
	SP	Poorly graded sands or gravelly sands, little or no fines	Good	Low	Low
	GM	Silty gravels, gravel-sand-silt mixtures	Good	Medium	Low
	SM	Silty sand, sand-silt mixtures	Good	Medium	Low
Group II	GC	Clayey gravels, gravel-sand-clay mixtures	Medium	Medium	Low
	SC	Clayey sands, sand-clay mixture	Medium	Medium	Low
	ML	Inorganic silts and very fine sands, rock flour, silty or clayey fine sands or clayey silts with slight plasticity	Medium	High	Low
	CL	Inorganic clays of low to medium plasticity, gravelly clays, sandy clays, silty clays, lean clays	Medium	Medium	Medium to Low
Group III	CH	Inorganic clays of high plasticity, fat clays	Poor	Medium	High
	MH	Inorganic silts, micaceous or diatomaceous fine sandy or silty soils, elastic silts	Poor	High	High
Group IV	OL	Organic silts and organic silty clays of low plasticity	Poor	Medium	Medium
	OH	Organic clays of medium to high plasticity, organic silts	Unsatisfactory	Medium	High
	Pt	Peat and other highly organic soils	Unsatisfactory	Medium	High

For SI: 1 inch = 25.4 mm.

a. The percolation rate for good drainage is over 4 inches per hour, medium drainage is 2 inches to 4 inches per hour, and poor is less than 2 inches per hour.

b. Soils with a low potential expansion typically have a plasticity index (PI) of 0 to 15, soils with a medium potential expansion have a PI of 10 to 35 and soils with a high potential expansion have a PI greater than 20.

oratory test results can be used to obtain more accurate bearing values to design the footings. The weight of various elements in a structure is determined by the structural designers, and the weights are carried to the ground as linear loads to strip footings or to spot supports, such as columns resting on spread footings, piers, or piles. Once these loads are determined, the foundation designer must design the footings to properly distribute the loads to the soil, sizing the footings to not exceed the allowable bearing pressures. The greater the loads the greater the area of the footing will have to be for the same strength soil. Figure 4.5 illustrates spread footings sized for different allowable pressures, but with same total load.

Soil bearing is determined by its strength. Strength can be determined by tests in the field and in the laboratory. Such tests are aimed at determining the resistance of the soil to being torn or sheared under the applied pressures. The field tests may consist of procedures done with drill-rig equipment, such as pushing a cone into the soil and measuring its resistance to penetration, using a "vane-in-hole" shear test, or driving a sampler into the ground using a known weight dropped a certain distance and counting the number of blows to achieve a certain penetration. Samples of soil recovered in the field may be used for estimating strength by squeezing with the fingers or, more accurately, using devices called a Pocket Penetrometer or a Tor-Vane. These are simple devices that can provide reasonably accurate determinations of the strength of the soil samples in the field. Such tests need to be carried out on undisturbed pieces of the soil, not ground-up material that might have been generated during the drilling process. The field and lab tests are discussed more fully in Chapters 6 and 7.

4.4 Compressibility

Compressibility is the property of a soil with a high water content that permits it to be reduced in volume by having the water squeezed out under the pressure of the foundation applied to it. Soils possessing such properties are tested in the laboratory, and not only the amount of settlement can be predicted but the time it may take to achieve such settlement. These are complicated tests and are subject to considerable errors ranging from disturbance of the sample in the field or laboratory to properly estimating the pressures at various depths and the locations of the drainage zones within the soil deposit. Usually silts or clays are the soils that can undergo significant compressibility. However, sands can be compressible if they have been deposited in a fairly loose condition. This can be tested by field procedures. Sands generally compress rapidly under applied loads while silts and clays compress slowly (sometimes over years).

A rule of thumb indicates that significant pressure increase can occur in the soil to a depth of about 85 percent of the width of the foundation element. A 24-inch-wide strip footing would exert increased pressures only for a few feet; however, a number of strip footings and spot footings combined over the area of the building may join up to increase the average pressures in the soil at greater depths permitting unexpected settlement to occur. Often a raft or mat-type foundation is used over compressible soils, and the depth of pressure increase is quite deep, being dependent on the total width of the mat, which is usually the width of the building. Thus it is possible for a small mat to increase pressures only within a shallow, low-compressibility soil zone. On the other hand, a larger building with the same foundation loading per square foot could increase

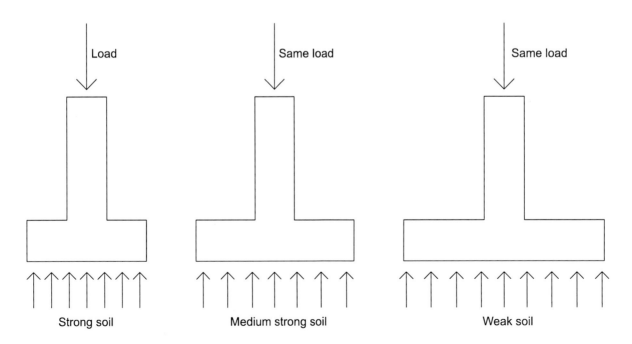

FIGURE 4.5
FOOTING SIZES VARY FOR SAME LOAD, DEPENDING ON SOIL STRENGTH

pressures within a lower, highly-compressible soil zone, producing large settlements. Figure 4.6 illustrates the depths of significant pressure increase due to the size of bearing width.

For residential construction, Section R401.4.2 of the 2012 IRC provides an alternate procedure if "compressible or shifting soil" is identified and can be removed and replaced with suitable soil. The determination of the depth required may be difficult without a geotechnical investigation if such soils are more than a few feet deep.

4.5 Volume-active Soils (Shrink and Swell)

Volume-active soils that undergo shrinking and swelling are frequently called "expansive clays." This type of soil is typically quite strong and will produce good allowable bearing pressures. However, depending on the location, the mineral composition of the soil, and its access to moisture variations, this soil can swell by taking on water and actually lift the building or individual footings enough to damage the structure. See Sections 1803.5.3 and 1808.6 of the 2012 IBC and Section R403.1.8 of the 2012 IRC. This type of soil has soil minerals that, because of internal microscopic structure, have the ability to expand from a dry state to a wet state. In effect the soil has a thirst for water and will absorb it if it is available. Figure 4.7 shows the approximate location of expansive soils in the contiguous United States. The map is a generalization, and other local areas may also have expansive soils.

Swelling of clay soils can be reduced or even prevented by sufficient pressure being applied to them to resist the tendency to expand. For this reason large and heavy structures are usually not troubled by expansive clay problems, except for lightly-loaded basement slabs. However, residential and light commercial structures on shallow foundations can be severely affected. Geographic locations are important because in many parts of the West and Southwest especially these climates tend to be arid or semi-arid with occasional wet periods. These conditions, combined with the right soil, can produce dramatic surface movements such as heaving or shrinking, which can range from a fraction of an inch to over 12 inches depending on the various conditions involved.

Sometimes trees are a factor in combination with high shrink-swell soils, causing unexpected foundation movement. The mechanism of the tree effect has been extensively studied and has two possible conditions. One condition is a tree located near a shallow foundation: this tree's root system removes moisture from under the foundation producing vertical shrinkage in the supporting soil. A second condition is the removal of trees from under the foundation during construction or from near the outside perimeter after construction. This allows the soil to regain moisture previously reduced by the root system, leading to localized expansion of the supporting soil. Figures 4.8 and 4.9 show the "tree effect."

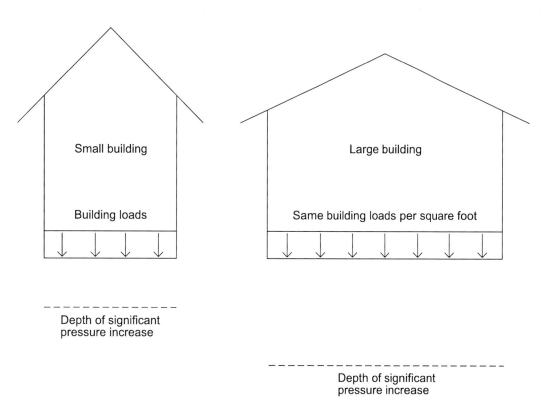

FIGURE 4.6
DEPTHS OF SIGNIFICANT PRESSURE INCREASE

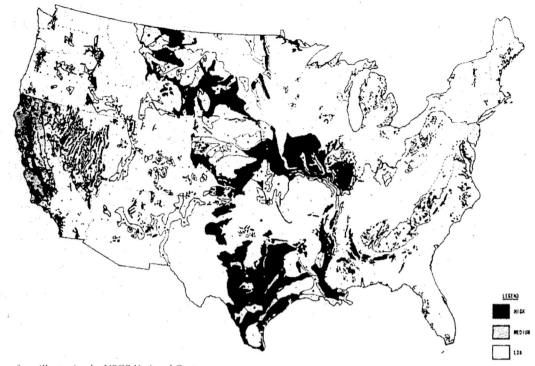

Taken from illustration by USGS National Center

FIGURE 4.7
DISTRIBUTION OF EXPANSIVE SOILS IN THE UNITED STATES

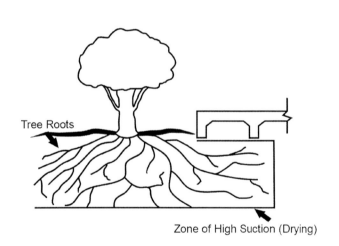

Tree Roots

Zone of High Suction (Drying)

FIGURE 4.8
VEGETATION EFFECTS ON FOUNDATIONS – TREE ROOTS EXTENDING UNDER FOUNDATION

THE SHRINKAGE OF CLAY SOILS WILL CAUSE THE FOUNDATION TO LOSE SUPPORT.

FIGURE 4.9
DRYING FROM TREE ROOTS CAUSING SOIL SHRINKAGE FORMING A SURFACE "BASIN"

A technique sometimes used to limit the tree root effect below shallow foundations is a root barrier. These barriers are 5 to 7 feet deep and are placed between the tree and the foundation about 8 to 10 feet each side of the tree. Figure 4.10 shows such a barrier.

4.6 Lateral Stability

A sloping site may have lateral stability problems ranging from gradual downhill creep to a major landslide. The steeper the slope the more likely such activity is to take place. Naturally, stronger soil or rock materials are less likely to move laterally than weaker ones. However, even stiff to hard clays can move laterally because of the cyclic change of moisture causing swelling and shrinking on a seasonal basis and softening of joints. This continual activity tends to move the upper soil in a downhill direction. Figures 4.11, 4.12, and 4.13 show indicators of downhill creep.

Significant cuts or fills must be retained by basement walls or retaining walls to achieve a vertical change in elevation. A clean material (clean means a soil with less than 5 percent passing the #200 sieve) such as a sand or gravel will not stand on a vertical cut, but will assume its natural "angle of repose," which is around 45 degrees or less. A stiff clay can stand vertically when cut even for a considerable height; however, it is unpredictable and may fail without warning because of softening in joints over time. This is one reason that trench shoring is so important for safety,

and measures must be taken to protect personnel entering trenches even in stiff clays.

If the site is laterally unstable and tends to move downhill, even if only fractions of an inch a year, a shallow foundation will be moved downward and will suffer distortion. Likewise, piers or piles can be pushed in a downhill direction and over time may even bend or break causing significant damage to the structure they support. Figure 4.14 illustrates the effect of downhill creep on piers or piles.

4.7 Highly Organic Soils

These soils have a large amount of organic or vegetable material mixed in with clay or silt. Their composition may be the result of vegetation caught up in floods and deposited with alluvium soils or peat bogs formed by meadows growing up in a wet area, often in depressions generated by past glacial activity. If the soils have a significant amount of organic materials, they usually are not suitable for any type of foundation support because of excessive settlement. Such soils will either need to be penetrated by deep piers or piles to a more stable material or totally removed and replaced with more stable material such as a gravel fill. Organic soils can be detected by an odor that is generated by decomposing vegetable matter. Comparing the liquid limit of oven-dried soil to that of the undried sample is a definitive procedure for detecting organic soil. See 7.3.3.3.

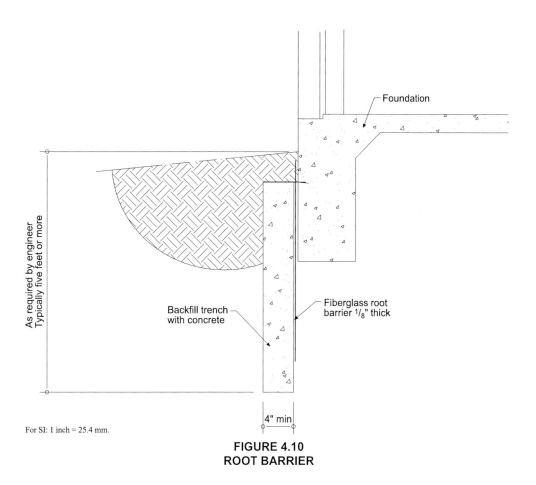

For SI: 1 inch = 25.4 mm.

**FIGURE 4.10
ROOT BARRIER**

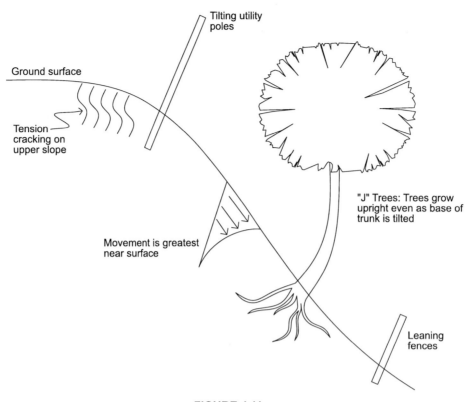

Tilting utility poles

Ground surface

Tension cracking on upper slope

Movement is greatest near surface

"J" Trees: Trees grow upright even as base of trunk is tilted

Leaning fences

FIGURE 4.11
EVIDENCE OF DOWNHILL CREEP

FIGURE 4.12
"J" TREE – AN INDICATOR OF DOWNHILL CREEP

FIGURE 4.13
PAVEMENT JOINT SEPARATION DUE TO LATERAL MOVEMENT CAUSED BY DOWNHILL CREEP

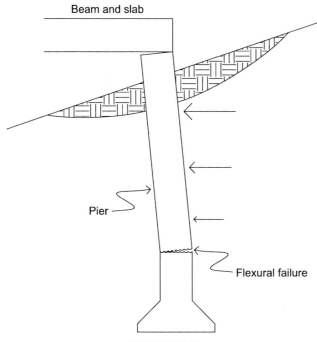

FIGURE 4.14
DOWNHILL CREEP AND PIER-SHAFT FAILURE

Organic deposits may also be found in man-made land-fills. These deposits are likewise unsuitable for foundation support. If there are considerable organics present either naturally occurring or in a former landfill, there also may be the generation of hydrogen sulfide and methane gas, which are dangerous both from toxicity and flammability. These are unsafe sites for constructed facilities.

4.8 Frost Heave

Different parts of the country are subject to freezing temperatures that can penetrate the earth from 6 inches to 6 feet depending on the climate. Frost heave, as the name implies, may cause foundations to be heaved up when ice is formed below the foundations. Pavements are sometimes subject to frost heave and can be destroyed by this action. It is important for the engineer to establish foundation elements below the frost line or use proper insulation arrangements. Frost depths are usually known in specific areas and may be mandated by the building department or can be determined by published literature for a locality. See Section 1809.5 of the 2012 IBC and Section R403.1.4.1 of the 2012 IRC. Table R301.2 (1), footnote "b" of the 2012 IRC requires the jurisdiction to fill in Frost Line Depth. Published literature from national or state engineering societies (such as American Society of Civil Engineers Standard 32) may assist in determining the frost line depth.

A special case of frost heave would occur in the case of a freezer storage facility in which freezing temperatures penetrate the foundation and go into the earth, causing frost heave or frost boils to occur. In these cases the depth of freezing temperature penetration is difficult to determine since the cold temperatures remain constant at the surface and are not seasonal. The published frost depth should not be relied on in these cases, and a special study is required.

4.9 TEST QUESTIONS

MULTIPLE CHOICE

1. Compressibility is the property of the soil with a high water content which permits it to be reduced in volume by (select one):
   a. trimming
   b. squeezing out water by loading
   c. compacting

2. Depressions in the soil surface generated by past glacial activity can result in the formation of (select one):
   a. boulder fields
   b. peat bogs
   c. earthquakes
   d. limestone

3. Contact of moving glacier ice with rocks can produce (select one):
   a. no discernible effect
   b. metamorphic rock
   c. major grinding action
   d. long-term cloud cover

4. Loess deposits often appear as stable vertical cliffs that (select one):
   a. remain stable even when saturated
   b. collapse when saturated
   c. always produce major dust storms

5. The USCS classification is dependent on (select one):
   a. how wet or how compact the soil is
   b. strength of soil
   c. grain size and Atterberg limits
   d. color

6. In the USCS all coarse-grain materials are larger than that which will pass through (select one):
   a. #4 sieve
   b. #200 sieve
   c. #100 sieve

7. In the USCS if the symbol carries the capital letter G, the material is (select one):
   a. granite
   b. clay (from the German word glaub)
   c. gravel

8. A highly organic soil is (select one):
   a. hardly ever a problem with regard to foundation establishment
   b. always is a problem with regard to foundation establishment
   c. always toxic to wildlife

9. A "rule of thumb" indicates that significant pressure increase can occur in the soil to a depth of about (select one):
    a. 15% of the width of the foundation element
    b. 50% of the width of the foundation element
    c. 85% of the width of the foundation element
    d. The longest dimension of the building

10. As a river or creek carries soil particles in suspension and as the velocity of the current slows, the first particles to be deposited will be (select one):
    a. silt
    b. organics
    c. gravel
    d. clay

11. Depending on the climate in a locality, freezing temperatures can penetrate the earth distances of (select one):
    a. 6 inches to 1 foot
    b. 6 inches to 2 feet
    c. 6 inches to 10 feet
    d. 6 inches to 6 feet

12. Footnote "b" of Table R301.3 (1) in the 2012 IRC requires a jurisdiction to fill in (select one):
    a. its address
    b. the name of the building official
    c. the frost-line depth
    d. the U.S. Weather Service website

Chapter 5

GROUNDWATER AND DRAINAGE

Major water problem due to seeping groundwater

Chapter 5

GROUNDWATER AND DRAINAGE

5.1 Types

Groundwater may be a deep-bottomed water table extending to considerable depths below the ground with the water surface at some position in the ground, or ground water may be a perched water table, which is a shallow-bottomed collection of water under the ground. Perched water is likely to be seasonal or intermittent water that occurs in cycles throughout the year depending on the previous rainfall.

5.1.1 Water table.

Water tables are found in many parts of the country at relatively shallow depths. Water is generally found in a porous material known as an "aquifer" and is supplied from some up-gradient area by rainfall, probably offsite from the construction being considered. Some water tables are recharged hundreds of miles away from the point where they are encountered. Wells for drinking water or agricultural purposes are established within water tables that are of sufficient volume and quality. The surface of the water table may fluctuate depending on supply and withdrawal. A number of major aquifers have been drawn down by overuse, rendering economical well production impossible. If a water table exists within a sloping terrain, it may outcrop at some point down slope, creating a seepage area or spring.

In many parts of the Gulf Coast (including Houston) and some places on the California coast, groundwater withdrawal for industrial or drinking water purposes has lowered the water table to the point that extensive regional settlement (also called subsidence) has taken place. If a water table exists at a certain depth, the soil below this depth has an effective weight of only about one-half of what it would above the water table. This is due to the buoyancy effect of the water in the soil. For example, a typical soil weighs 120 pounds per cubic foot. Water weighs 62.4 pounds per cubic foot. If soil is below a water table, the unit weight would be 120 minus 62.4 or 57.6 pounds per cubic foot. When the water table is lowered, this buoyancy effect is removed, and the soil effectively doubles its weight and may cause consolidation of the general soil profile. Regional subsidence in Houston has been documented at over 9 feet in places. This effect depends not only on the amount of the water table reduction but also the properties of the soil.

5.1.2 Perched water.

Perched water is similar to a water table except that it does not extend to great depths. For example, it could be a layer of gravel extending from the ground surface down 10 feet, resting on a clay or hard-pan type material. Lowering of perched water tables will generally not cause the regional subsidence associated with draw-down of a major water table as described above, but perched water tables can be difficult for construction purposes. Because they are usually shallow and of small extent, they are often seasonal. An excavation made in a dry period may not find a perched water table, but after a wet season there may be water near or exiting the ground surface.

5.1.3 Seasonal or intermittent water.

Water may appear near the surface or outcropping on the surface in the form of springs or seepage after prolonged wet periods and then disappear after the region dries out. This could be a perched water table described above or perhaps simply shallow soil that holds water for a period of time and then loses it during drying periods. Perched water and seasonal or intermittent water are frequently the same thing. One problem with seasonal or intermittent water is that a site may appear dry in test borings placed on the site but later develop water during construction or during occupancy when it is least expected. For these reasons the foundation designer should consider the possibility of seasonal water and specify waterproofing and drainage appropriate to keep water out of the below-grade habitable areas of a building.

5.2 Negative Effects of Uncontrolled Groundwater

Groundwater can be a problem during foundation construction. Drilled piers encountering water will likely have to be cased or otherwise processed to provide a relatively clean and dry hole for concrete and steel placement. This greatly adds to the expense of construction. Basements or split-level construction may be affected by groundwater, and provisions must be made during and after construction to keep the excavation and later the finished spaces dry.

Groundwater on a construction site may require temporary site-water protection and drying ranging from sheet piling and pumps to diversion ditches or even the addition of lime to dry a clay subgrade so construction can proceed. Uncontrolled groundwater may appear on the surface after the conclusion of construction causing a wet seepage area to appear part of the year, which can cause foundation or pavement failure or just be irritating when it impacts people's yards or other landscaping. It is reasonable to assume that groundwater may appear for only a short time during rains at sufficient elevations to get into below-grade areas, such as basements or split-level structures. See Section 1805 of the 2012 IBC and Sections R405 and R406 of the 2012 IRC for code information about foundation drainage, dampproofing, and waterproofing. Section 1805.3 discusses waterproofing of floors and walls when a hydrostatic condition exists. Even though permitted by the code, waterproofing without a ground-water control system is likely to fail since even a small gap in the waterproofing membrane can permit water penetration. If the ground water can rise above the finished floor level, the situation is similar to that of a boat hull in the water; a boat is carefully built to be waterproof, but bilge pumps are still provided.

Groundwater on a construction site that has had significant earthwork modification may outcrop in unexpected areas and lead to not only muddy seepage zones but possibly landslides that were not anticipated.

5.3 Site Water Problems

5.3.1 Evidence of problems.
Groundwater problems can frequently be anticipated by local experience, test borings, or simply observed during construction excavation, which is the worst time to find them because of the impact on the contractor's schedule and costs of construction. Indirect evidence includes certain types of vegetation, such as bamboo or willows, growing in particular areas where they are not normally found. This may indicate a seepage area. Local experience may indicate that certain areas are prone to springs or seeps, either intermittent or continuing. Older place names such as "Spring Hill" or "Pond Road" may be clues. All these indicators should be considered when evaluating a potential for groundwater. Test borings may reveal groundwater or the potential for seasonal water due to the types of soils encountered, but if the borings are drilled during the dry season, they may not be 100 percent reliable. For major projects, long-term readings of water levels in monitoring wells (also called piezometers) are frequently used to get the full picture.

5.3.2 Surface-drainage techniques.
Water that is on the surface, such as rainfall runoff, needs to be controlled in the design and construction of all structures. If it is not controlled it can create erosion, or it can pond adjacent to buildings, softening foundations or causing heave of swelling clays. In addition, if surface water is not cared for properly it may accumulate during rainstorms to the point where it penetrates into the structure causing problems for the occupants and physical damage inside the building.

Techniques for dealing with surface drainage include grading the site, providing a satisfactory protective backslope around the perimeter of structures, or providing drainage systems such as inlets and pipes. To properly design surface drainage, some estimate of the quantity of flow must be made for the various local areas, and proper drainage features must be provided. A good rule of thumb to protect structures such as residences and commercial buildings would be to slope the surrounding soil areas away from the building at about 6 inches in 10 feet (or 5 percent) away from the building. Hard surfaces, such as concrete, sidewalks, or patios may be sloped at less than 5 percent. Typically, a 1/8-inch to 1/4-inch fall in 12 inches is adequate (between 1 percent and 2 percent) for hard surfaces. The general site should be drained with a minimum of 1 percent or 2 percent slope (1 to 2 foot fall per 100 feet) using continuously sloping swales to convey the water that may fall on the site or come off the roofs safely away to a street or other drainage facility. Overland flow of off-site storm water may also need to be considered in an overall drainage plan. Surface drainage is discussed in 2012 IBC Section 1804.3 and in 2012 IRC Section R401.3.

It is also a good idea to provide a proper "reveal" around foundations, typically placing the finished floor 6 to 8 inches above the adjacent ground surface. It is important to prevent the ground surface being built up to an unacceptable level due to later additions of topsoil or landscaping features. Figure 5.1 illustrates a typical drainage and grading plan for residential construction.

5.3.3 Sub-surface drainage techniques.
If sub-surface water is a problem, drainage must be provided to remove the water down to a specified level within the soil. This is frequently done with something called a "French drain," which is also called a sub-surface drain. These drains consist of sand or gravel placed in a trench, usually draining to a slotted or perforated pipe in the bottom of the trench. This pipe will collect the water and draw the surface of the water down to the level of the pipe if it is properly designed and constructed. The pipe and trench bottom should be sloped at a minimum of 0.5 percent, which is a 6 inch fall in 100 feet, to provide positive drainage. A slope of 1 percent is preferable if feasible. Sub-surface drains' trench bottoms should not be permitted to be constructed with sags or "bellies" that will collect water and hold it for an extended period of time. This may defeat the purpose of the subsurface drain. The perforated drain pipe should rest directly on the trench bottom, not over several inches of sand or gravel, to avoid the reservoir effect.

Walls of basements or split-level construction should be protected with an adequate drain as described above, with the pipe bottom at least 6 inches below the floor to be protected, using proper pipe slope. In addition, occupied areas should be protected with a waterproofing material on the outside of the walls. Waterproofing alone generally cannot prevent all water penetration if a significant quantity of water collects at elevations above the finished floor. It usually takes a combination of sub-surface drains and waterproofing to properly protect habitable areas. All subsurface drains should continuously slope to a "daylight" discharge. Running sub-surface drains to a so called "dry well" is rarely effective and frequently makes the problem worse. Sump pumps should only be used as a last resort, since such pumps can fail unnoticed for long periods of time. On the other hand, gravity drainage is constant and requires no maintenance or monitoring.

Sometimes it is important to keep moisture from penetrating underneath foundations, either because capillary action or vapor rise in the foundation generates moisture within the structure or because there are expansive clays underneath the foundation that should be protected from water increase. In these cases often a "barrier drain" is employed. This consists of a French drain as described above with the addition of a waterproofing membrane up the side of the protected foundation, wrapping under the drain pipe. Any water that passes toward the sub-surface drain would be collected and carried away and none would be permitted to pass through the adjacent soil wall barrier, even in vapor form. Figures 5.2 and 5.3 illustrate typical sections of sub-surface drains and barrier drains. Figure 5.4 illustrates a typical detail for protecting below-grade habitable areas.

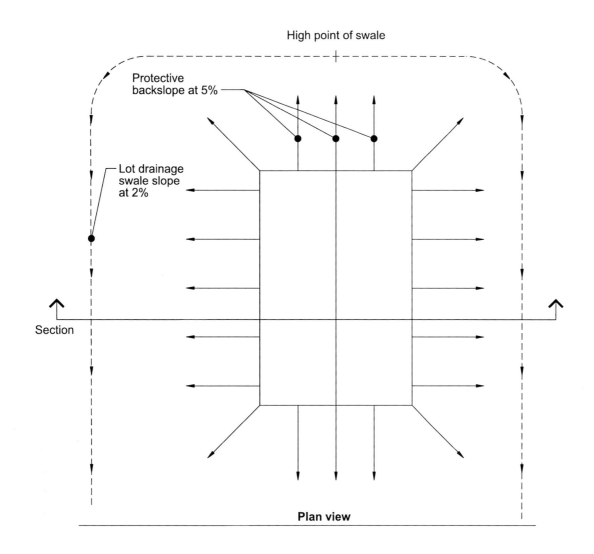

High point of swale

Protective
backslope at 5%

Lot drainage
swale slope
at 2%

Section

Plan view

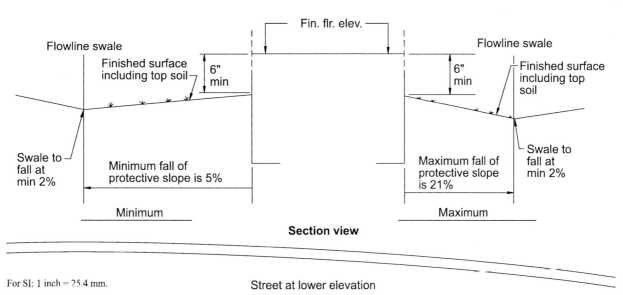

Fin. flr. elev.

Flowline swale

Finished surface
including top soil

6"
min

6"
min

Flowline swale

Finished surface
including top
soil

Swale to
fall at
min 2%

Minimum fall of
protective slope is 5%

Maximum fall of
protective slope
is 21%

Swale to
fall at
min 2%

Minimum

Maximum

Section view

Street at lower elevation

For SI: 1 inch = 25.4 mm.

FIGURE 5.1
EXAMPLE OF TYPICAL DRAINAGE PROVISIONS FOR A RESIDENCE

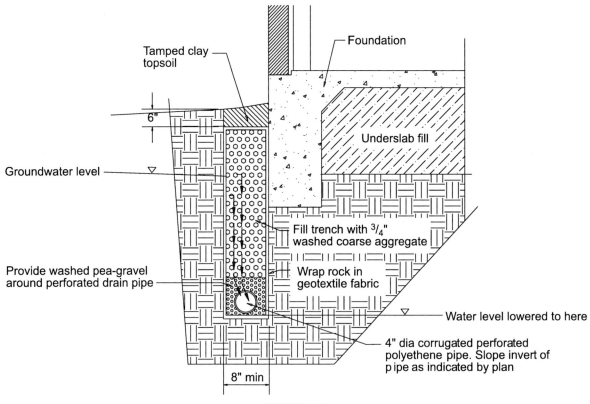

Tamped clay topsoil

Foundation

6"

Underslab fill

Groundwater level

Fill trench with ³/₄" washed coarse aggregate

Provide washed pea-gravel around perforated drain pipe

Wrap rock in geotextile fabric

Water level lowered to here

4" dia corrugated perforated polyethene pipe. Slope invert of pipe as indicated by plan

8" min

FIGURE 5.2
SUBDRAIN DETAIL

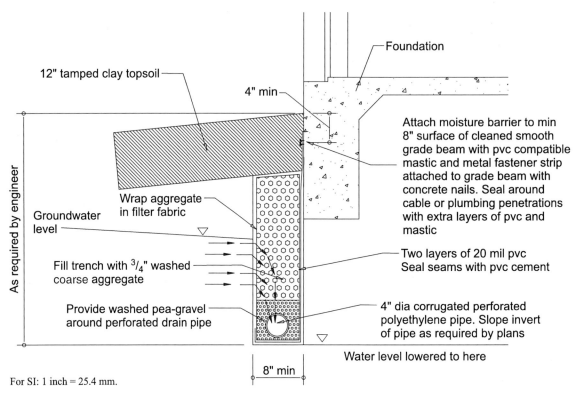

12" tamped clay topsoil

Foundation

4" min

As required by engineer

Wrap aggregate in filter fabric

Groundwater level

Attach moisture barrier to min 8" surface of cleaned smooth grade beam with pvc compatible mastic and metal fastener strip attached to grade beam with concrete nails. Seal around cable or plumbing penetrations with extra layers of pvc and mastic

Fill trench with ³/₄" washed coarse aggregate

Two layers of 20 mil pvc Seal seams with pvc cement

Provide washed pea-gravel around perforated drain pipe

4" dia corrugated perforated polyethylene pipe. Slope invert of pipe as required by plans

Water level lowered to here

8" min

For SI: 1 inch = 25.4 mm.

FIGURE 5.3
BARRIER AND SUBDRAIN

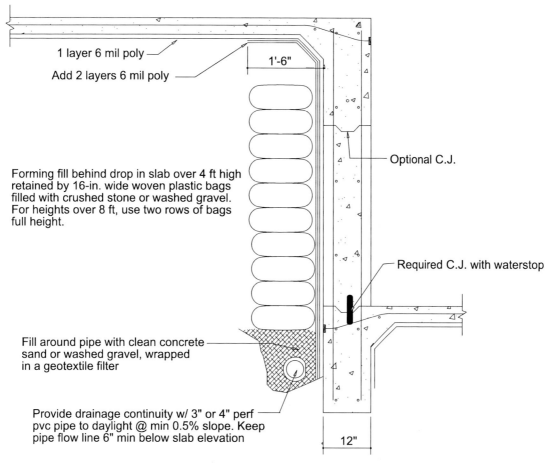

1 layer 6 mil poly

Add 2 layers 6 mil poly

1'-6"

Forming fill behind drop in slab over 4 ft high retained by 16-in. wide woven plastic bags filled with crushed stone or washed gravel. For heights over 8 ft, use two rows of bags full height.

Optional C.J.

Required C.J. with waterstop

Fill around pipe with clean concrete sand or washed gravel, wrapped in a geotextile filter

Provide drainage continuity w/ 3" or 4" perf pvc pipe to daylight @ min 0.5% slope. Keep pipe flow line 6" min below slab elevation

12"

For SI: 1 inch = 25.4 mm, 1 foot = 304.8 mm.

FIGURE 5.4
PROTECTION FROM WATER PENETRATION OF HABITABLE AREAS BELOW GRADE

Section 1805 of the 2012 IBC and Section R403.1.7.3 and R405 of the 2012 IRC discuss foundation drainage requirements. It is important to be sure there is at least a 0.5 percent slope on all drainage conveyances, including the subgrade below the floor base, to prevent the conveyances from holding water within the drainage system.

5.4 TEST QUESTIONS

MULTIPLE CHOICE

1. Three bad results for a construction project resulting from uncontrolled runoff of surface water are (select three):
   a. contractor and inspector disagreements
   b. problems with piling
   c. erosion
   d. flooding into structure
   e. heave of swelling clays

2. If the subsurface condition at a site appears dry when placing test borings (select one):
   a. there will never be a groundwater problem at this locality
   b. the borings were in error
   c. groundwater may appear later
   d. groundwater is never a problem

3. To ensure that below-grade habitable areas of a building are not affected by groundwater penetration it is necessary to (select one):
   a. ensure good vibration of concrete during construction
   b. seal the upper ground surface
   c. provide waterproofing and water-control drains
   d. use two coats of paint

4. The best time to find groundwater is (select one):
   a. during construction excavation
   b. before construction begins
   c. at the end of construction

5. The surface of a water table may fluctuate depending on (select two):
   a. weather cycles (supply)
   b. groundwater pumping (withdrawal)
   c. alignment of the moon
   d. wildlife density

6. What is the net unit weight of a typical submerged soil as discussed in an example in this book? (select one):
   a. 20 PCF
   b. 150 PCF
   c. 57.6 PCF
   d. 120 PCF

7. Continuously sloping swales to convey surface water from a site should be sloped at a minimum of (select one):
   a. 8 percent
   b. 15 percent
   c. 1 percent
   d. 25 percent

Chapter 6
SITE INVESTIGATIONS

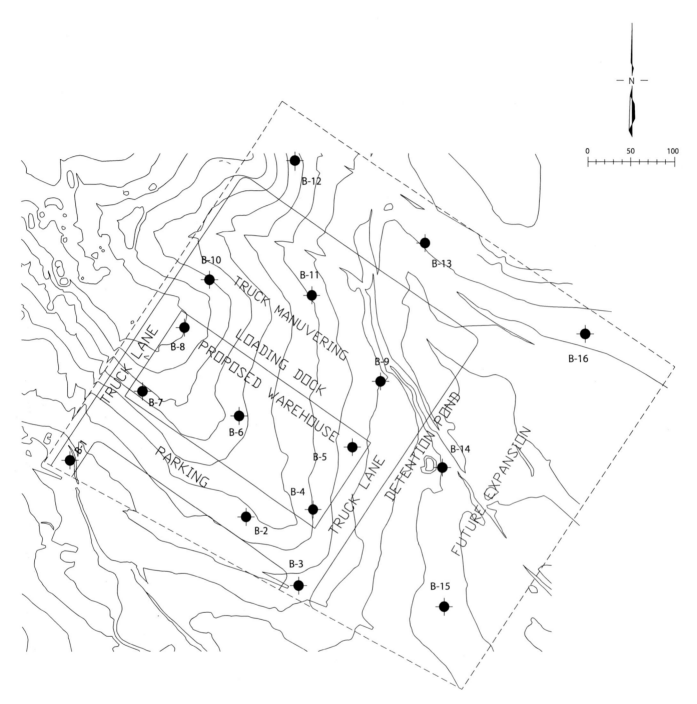

A borings plan

Chapter 6

SITE INVESTIGATIONS

6.1 Purpose

A site investigation is the prelude to rational foundation design for any structure including buildings, bridges, dams, retaining walls, pavements, or any other constructed facility that must eventually take its loads to the earth or resist earth loads; see Section 1803 of the 2012 IBC and Section R401.4 of the 2012 IRC. A site investigation report can convey the nature of the site below the ground surface along with some insights concerning the visible surface to the designers of the structure and the contractors who will build it. The factors that would impact the design or construction activities can be clarified and documented in a site investigation report. This report, also known as a "geotechnical report" or a "soils report," should indicate the soil or rock strata underlying the site and provide information about their properties for engineering purposes. Other information should include recommendations for foundation types, depths of establishment, safe bearing pressures, groundwater conditions, and insights into constructability. For all important engineering work, the foundation cannot be rationally designed without some type of guidance from the site investigation report. Without it the designer simply must guess. This chapter presents a somewhat detailed discussion of site investigations, since the quality of the investigation report cannot be better than the quality of the field data and sampling.

6.2 Preliminary Studies

Often people in the architectural, engineering, or construction professions say, "We need to get some test holes drilled." A proper investigation includes a number of preliminary studies prior to drilling the test holes. Preliminary research from published information or from the geotechnical laboratory's experience and job files in the area may include likely geologic formations to be encountered; a study of the land forms and topography; a study of the vegetation, rock outcrops, and visible seepage areas on the surface; notation of the presence of fills, landfills, abandoned foundations, or underground tanks; and a study of aerial photographs. Additional information can be obtained from the U.S. Department of Agriculture's Natural Resources Conservation Service county soil reports. These reports typically describe near-surface soil profiles to depths of 3 to 6 feet. The preliminary site-development plan, topographic maps, and existing utility locations should also be assembled.

6.3 Planning the Investigation

With the preliminary studies done, the geotechnical engineer will need to know more detail about the proposed building's size, the type of building, and the approximate gravity, wind, and seismic loads that need to be dealt with in the foundation recommendations. Once this information is digested, the geotechnical engineer can lay out the boring pattern, establish the depth of borings, and determine the type and quality of sampling and field testing that might be necessary. To develop a budget for the client, the engineer also needs to estimate the type and number of laboratory tests.

Generally, the client's budget factors in at this point, and a great deal of pressure might be applied to do the site investigation for the least amount of money. For the engineer to yield to this pressure is generally bad economics for two reasons. First, if the geotechnical engineer has to work with a limited scope of field and laboratory work, the information obtained is much less informative, and the recommendations must be suitably conservative to cover the missing data that might have permitted a more precise analysis. Overly conservative recommendations translate into higher construction costs. Second, when an insufficient investigation is done, important features of the site may be missed and important soil properties may not be properly determined, leading to a failure of the foundations or earthwork after construction.

Another factor that the engineer should consider while planning the investigation is the type of equipment that may be necessary to access the various test locations in the field. Typically, truck-mounted drilling equipment is used, but in some cases, terrain or the trafficability of the site is such that tracked vehicles or swamp-buggy type equipment may be necessary to reach the locations of the various test borings. In other cases borings are done on water from a barge; investigations for bridges or for building structures that are to be built over water are examples. There are many ways to obtain field information other than a typical coring rig. A foundation pier drilling rig can be used, or test pits can be excavated using backhoes, bulldozers, or draglines depending on the site situation. The object is not just to make holes in the earth, but to obtain information and samples in an orderly fashion.

After the investigation is planned and the scope established, typically the client would like to know an approximate range of the budget and sometimes an exact figure. The client should be advised that field conditions or soil properties may be encountered that were not anticipated and the budget might need to be amended. This information is presented to the client and an authorization is obtained. It is important for the geotechnical engineer to be assured that he or she can have legal access to the site and won't be shot at by an angry farmer. Accurate site location information is necessary, since test borings have been known to be drilled on the wrong site.

6.4 Field Procedures

To begin the field process, the drill crew is provided a map showing the location of the site and proposed borings, and the crew is given specifications for the various test holes, such as depth, type of samples, and field testing that will be required. Depending on the size and importance of the project, often a geotechnical engineer will accompany the field crew to oversee the process; however, this is not universally done, especially for smaller projects. Boring locations are typically laid out in the field by taping or approximate survey methods by the field crew. For important projects it may be necessary to employ a professional surveyor to precisely locate the borings and to determine the ground surface elevations for future relationship to the project plans. GPS equipment is often used to verify locations by the field crew, possibly followed up by a surveyor to get exact locations. Sometimes it is not possible to access the original proposed locations of each boring because of field conditions. If the locations are moved even a minor amount, the new locations should be noted. If a large relocation is needed, the geotechnical engineer should be consulted for possible re-design of the investigative program or a change of equipment. The geotechnical engineer should also be consulted if unexpected subsurface conditions are discovered.

The large majority of site investigations are done with a drill truck (also called a core rig), which is equipped with lengths of drill pipe or continuous flight augers to penetrate the earth. Some investigations in unstable soil require the use of "hollow- stem" augers, which are augers that are drilled into the earth in sequence, adding 5 to 10 feet to the drill string, providing an open-cased hole as the drilling proceeds downward. The hollow-stem auger has a 4-inch to 8-inch hollow interior, permitting samplers to be inserted and withdrawn at the bottom of the drilled hole as depth is advanced.

Some drill equipment uses a rotary bit attached to drill pipe with circulating wash water or compressed air to remove the cuttings to the ground surface. This technique is almost universally used if hard materials must be penetrated or rock must be cored. Rock coring is done with a double-tube core barrel with tungsten carbide or diamond inserts in the annular drill bit. The double-tube arrangement permits the outer tube to rotate while the inner tube remains stationary, protecting the sample from being disturbed by rotation, the return wash water, or air. Figures 6.1 through 6.8 are photographs of various kinds of exploratory and sampling equipment.

FIGURE 6.1
PORTABLE DRILL RIG

FIGURE 6.2
TRUCK-MOUNTED DRILL RIG
FOR SITE INVESTIGATION

FIGURE 6.3
BACKHOE AND TEST PIT

FIGURE 6.6
ROTARY WASH BIT

FIGURE 6.4
SOLID AND HOLLOW-STEM CONTINUOUS AUGERS

FIGURE 6.7
DOUBLE-TUBE CORE BARREL DISASSEMBLED

FIGURE 6.5
TUNGSTEN CARBIDE ROCK BIT SOLID-STEM AUGER

FIGURE 6.8
ROCK CORES OBTAINED WITH CORE BARREL

6.5 Sampling

Sampling may be done by taking grab samples of auger cuttings, which is a fast and inexpensive way to obtain disturbed samples of the soil for further examination. More sophisticated sampling includes the use of sharpened thin-wall tube samplers (also known as Shelby tube samplers), which are pressed into the ground like a cookie cutter for a distance of 6 inches to 24 inches, depending on the soil materials. Samples recovered in this manner are usually extruded from the tubes in the field, although sometimes the tubes are simply capped and carried to the laboratory for extrusion there.

Other types of sampling equipment include a thick-walled split-spoon sampler, which is smaller than the thin-wall tube sampler and is driven by a drop hammer. This sampler is used in sands and small gravels that are not suitable for the thin-wall tube sampler. The sample is caught inside this sampler, but it is significantly disturbed. The number of hammer blows required to advance the sampler are recorded for each 6 inches until 18 inches has been driven. The last two sets of 6-inch drives are added together, and the number of blows required to penetrate this foot is called the "N value." The N value has been correlated with different types of formations and can be a useful indicator of strength and settlement potential.

There are many other types of sampling equipment, and some are quite sophisticated. Their purpose is to get samples from the field that are suitable for the laboratory tests that are scheduled to be run. Some samples can be disturbed and still provide the information needed, while other samples need to be nearly undisturbed and represent the structure and condition of the soil in place. All undisturbed field samples must be wrapped and sealed against moisture-content change, secured against distortion and heat, and carefully transported to the laboratory quickly. If undisturbed samples are left in sampling tubes for transport to the laboratory, the ends of the tube should be sealed with micro-crystalline wax and the whole tube enclosed in plastic wrap. Auger grab samples may be placed in sealed bags to avoid moisture change. All samples must be marked with clear identification. Figures 6.9 and 6.10 are photographs of samplers.

Other site observations that the field crew might make would be to extend the preliminary studies by field observations noted by sketching items of interest on the field boring location plan.

6.6 Logging

While advancing the hole, the drill crew or geotechnical engineer will produce a log of the boring. A typical boring log will indicate the different types of soil strata and at what depths encountered, where samples were taken and types of samples, a description of the soil based on field observations, and a record of any type of field testing done. Other information may include the depth at which groundwater was first encountered and the level to which it stabi-

FIGURE 6.9
SHELBY TUBE SAMPLER USED FOR FINE-GRAIN SOILS

FIGURE 6.10
SPLIT-SPOON SAMPLER USED FOR SANDY SOILS

lized after a certain period of time. It is also sometimes useful to make a record of the difficulty in drilling or the difficulty in keeping the hole open, because this may add information useful to the design or construction of the project. An experienced logger in the field can give a reasonably good description of the soil as it is penetrated and sampled using field expedient tests and experience. The presence of organics and unusual odors should also be noted. The geotechnical engineer adjusts field logs in the office, using more accurate laboratory testing and classification, and finished logs are prepared for the investigative report. Figure 6.11 is a photograph of a geotechnical technician keeping the field log.

FIGURE 6.11
KEEPING THE FIELD LOG

6.8 Field Tests for Groundwater

Test-boring logs routinely note if groundwater is encountered during the drilling. Usually the first depth of encounter is noted, and at the conclusion of the boring another observation is made and the time and depth is noted. In some cases where groundwater can be a significant problem, it is possible to convert the test-boring hole into a groundwater monitoring well for semi-permanent observations over an extended period of time. Such monitor wells are constructed with sand around a pipe with slits to permit water to enter the pipe in the ground, much like a water well might be installed. A surface casing is extended to the top of the ground and is typically fitted with a locking cap. Water levels can be observed periodically, usually by lowering a calibrated pair of wires with contacts at the end that will complete an electric circuit when they encounter the water surface. Other more sophisticated methods can also be employed, including automatic remote sensing devices. Pumping and draw-down tests can also be performed on monitor wells, which can be used to estimate the quantity of water that may be entering an excavation. Discussion of such procedures is beyond the scope of this book.

6.9 Field Tests for Constructability

The field crew should note conditions of the test boring as it is advanced, such as where the hole caves and the presence of trash, boulder deposits, and groundwater. These observations can be made with the standard 4- to 6-inch diameter test boring, but sometimes the information is imperfect. On major jobs in which drilled piers or piles may be utilized, it is sometimes a good idea for the engineer to use a large hole diameter foundation drill rig and

observe the drilling conditions in order to provide designers and contractors with an estimate of whether or not the holes will have to be cased during construction and other factors that could affect the bid prices of the piers or piles. One particular procedure of interest is the construction of an under-ream or "bell" at the bottom of a drilled pier. Since this is an undercut at the bottom of the pier shaft, caving soil conditions can drastically affect the ability to place belled foundations in such locations.

Other constructability information that could be obtained during the field investigation includes the nature of rock deposits found such as their hardness and whether or not they are layered. This information can significantly affect the cost of excavating rock. Figures 6.12 and 6.13 illustrate a typical field-log form and a finished report-ready boring log.

6.9 Other Site Observations

Other observations may include the presence of springs, seeps, unstable land forms, fill, landfills or dumps, trees (whether second growth or original growth), areas where trees may have been removed, old trails and roads, and fence lines. They may also include indications of previous construction on the site, such as old foundations, buried tanks, septic tanks, and drain fields. Sometimes the ability of the drill truck to drive around the site is noted, as it may be an important indicator. It is a good idea for the engineer or crew to take a number of photographs of the site during the field investigation. Field tests that can be performed other than the "blow count" procedure described in 6.5 could include a vane-in-hole shear test, in which the amount of shearing torque required to rotate a specified vane device is recorded as an indicator of shear strength. Other tests that are sometimes done include cone penetrometer tests—either driven by drop hammers or pressed continuously into the soil—to develop a profile of shear strength and water pressures. Samples obtained in the field are typically tested for an estimate of the shear strength using a Pocket Penetrometer or a small hand held Tor-vane; see Figure 7.13.

Sometimes useful information concerning the potential of a site for settlement can be obtained by the engineer's observation of the condition of the nearby structures and pavements. Major structures often have deep basements of multiple stories below grade. Information should be presented in the site investigation report concerning the effect of excavation on nearby structures, pavements, and other infrastructure.

The presence of active or inactive faults should be studied through the office preliminary studies and supported by field observations. Mapping significant faulting will probably require the services of a specialist, such as a geologist or specifically experienced geotechnical engineer. Active faults can be damaging to structures and must be noted and made part of the record of the site.

LOG OF BORING

Project : _____   Project No.: _____

Boring No. : **B-3**   Logged By: **RJB**   Rig Used: **B-53**   Project City: _____

Date: _____   G.S.E.: _____

Start on Hole: **8:30 AM**   Finish on Hole: **1:00 AM**

DEPTH	SAMPLE	CORE	TYPE OF SOIL	MATERIAL DESCRIPTION	Fill?	Color	Pocket Pen (Clayey)	JTD (Clayey)	Shine (Clayey)	Quantity (Sandy or Gravelly)	Size (Sandy or Gravelly)	Moisture Condition	Symbol (Unified or Geologic)	Blow Count	Other
0				0.0 DK GRAY/DK BRN, HD, CLAY, TRACE OF ROOTS	NO		3.75			—		1/2	CH		
	2.5			1.5 NO ROOTS			4.5			—		2	CH		
	4.0			2.5 DK RED TO RED											
5				4.0 LT. TAN, CALC, TRACE OF SAND/GRAVEL			3.25			I		2	CL		
	6.0			PUSH REFUSAL @ 7.5'			3.5			2		2	CL		
	7.5			PUSH REFUSAL @ 8.5'			4.5			2	1/4"	2	CL/SC		100
	8.5														
10	10			10.0 HEAVY GRAVEL, W/ YT CLAY & SAND,			4.5+			5	1/2"	3	GC	2/5/6	11.5
	11.5			''' WATER ENCOUNTERED @ 10.5'											
	13			12.2 YT, TAN, LT, GRAY, STIFF, CALC,			3.5			—		2	CL/CH		
15	14.5			13.0 RED, REDDISH BRN SEAMS											
	16			14.7 V. STIFF, TO HARD (LS?)											
	17.5			15.1 MED, STIFF, BLACK SHALEY,			4.5			—		2	CH		
20				17.8 SHELBY TUBE REFUSAL			3.75			—		2	CH		
				19.0 STIFF, TO HARD (LS)			4.5			—		2	CH		
				20.0 STIFF, TO MED STIFF (SHALE), V STIFF TO V. STIFF LYRS			4.5+			—		1	CH		
25	25.0			23.5 V. HARD, TAN, LS,			—			—		—	LS		
				25.0 START CORING (WATER ADDED)											
				26.6 W/CLAY LYR, PARTIAL WATER LOSS											
30				SCR = 49" RQD = 39" TD = 30.0'											

RUN No.: 49"/39"

Sandy or gravelly code:

| 1 = Trace | 2 = 1/10 | 3 = 1/4 |
| 4 = 1/2 | 5 = 3/4 | 6 = all |

Moisture Condition code:

| 1 = dry | 2 = damp | 3 = moist |
| 4 = wet | 5 = water entering hole |

Coring Code:

Recovery in./in.
RUN No.: # 48/60

6.0 — Start of RUN

11.0 — End of RUN

Soil Sampling Code:

■ Push Tube   ∫ Auger

⊠ Split Spoon   ▯ Poor Boy

Write NR by symbol if no sample recovered.

Blow count per:

XX
XX   6"
XX

Check as applicable:

☑ Auger used   Size **4.5"**   ☐ Air Used

☑ Drill water used   Bit Size **4.5'**

☐ Hole caved at _____ ft.   Noted after _____ hrs.

☑ Hole filled   ☐ Hole left open

Blow count by: (SPT)   ☐ THD Cone

FIGURE 6.12
FIELD LOG

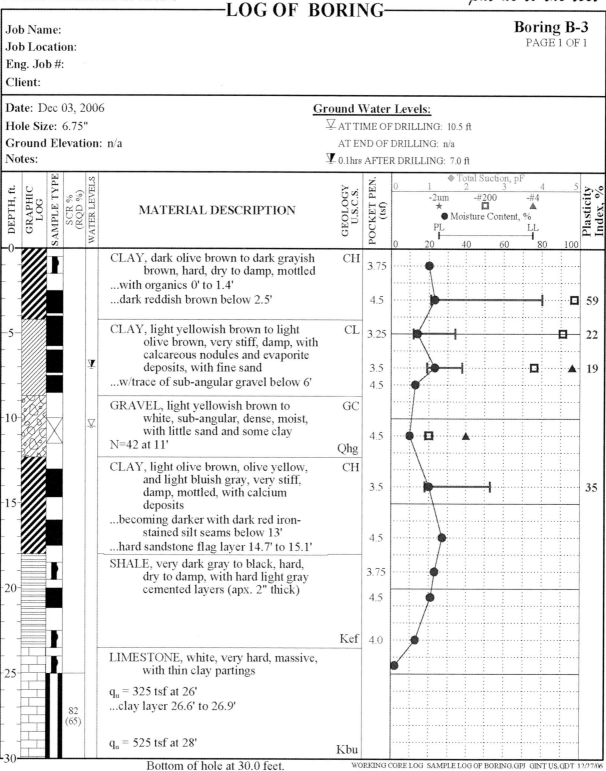

**FIGURE 6.13
REPORT-READY BORING LOG**

6.10 TEST QUESTIONS

MULTIPLE CHOICE

1. Two reasons why an usually low site-investigation budget will not be a good idea (select two):
   a. overly conservative recommendations
   b. no pavement recommendations
   c. failure of foundation
   d. to save money for developers

2. A geotechnical report should provide recommendations for (select two):
   a. landscaping soils
   b. foundation types
   c. safe soil bearing pressures
   d. locations for organics disposal

3. Studies preliminary to performing a site investigation include assembling the following (select three):
   a. site-development plan
   b. prevailing wind patterns
   c. topographic maps
   d. existing utility locations
   e. available magnetic surveys

4. The likely geologic formations that may be encountered (select one):
   a. play no part in preliminary studies for a site investigation
   b. are only of interest in the final analysis of the geotechnical report
   c. are important for preliminary planning
   d. will always eliminate the need for further site investigations

5. To lay out the boring pattern, the geotechnical engineer will need to know about (select two):
   a. proposed building size and type
   b. most convenient locations for borings
   c. building loads
   d. available legal access

6. The main objective of the site borings is (select one):
   a. completion of field work within budget
   b. to meet the time schedule
   c. obtain information and samples in an organized fashion

7. Truck-mounted drilling equipment is (select one):
   a. the only way to investigate a site
   b. one of several methods to investigate a site
   c. should always be used since the client expects to see it

8. Hollow-stem augers are used to (select one):
   a. obtain a sonic reading of the subsurface conditions
   b. get samples on sites with unstable soil conditions
   c. obtain a more complete record of groundwater

9. The Shelby tube sampler is (select one):
   a. useful in sands and small gravels
   b. useful in clays and silts
   c. a device that can cut rock

10. A boring log is (select two):
    a. a piece of buried timber encountered by the drill crew in the field
    b. a record of the travel time on the site investigation
    c. a record of soil strata encountered

11. The presence of active or inactive faults (select one):
    a. is not of interest during the site investigation since these items can be handled during construction
    b. may require mapping by a geotechnical specialist
    c. is impossible to determine during a routine site investigation
    d. is shown in a record of samples

Chapter 7

TESTING ROCK AND SOIL

"One test is worth a thousand expert opinions."

Chapter 7

TESTING ROCK AND SOIL

7.1 Purpose

Rock and soil can be tested in the field and in the laboratory to determine the engineering properties of the materials that may be important for determining the depth of footing establishment, the type of footing, and the allowable bearing pressures. The purpose of site investigations is to provide information for the design team and the contractor for the construction of the facility, but even the most carefully performed investigation may still miss some hidden surprises. This is because the actual subsurface investigation is typically done with fairly widely spaced boring locations, which can be anywhere from 20 feet to 200 feet apart depending on the needs of the project. The investigation provides numerical values for the designers to use, based on field observations and laboratory tests, with a large dose of the geotechnical engineer's experience and judgment thrown in.

7.2 Laboratory Testing of Rock

7.2.1 Strength. The strength of rock cores is typically determined by a crushing or compression test in the laboratory. In this test the segment of the rock core selected for testing is cut with a diamond-blade rock saw to produce parallel and square ends of the core. The core is then capped with a rubber cushion or a high-strength compound and tested in a compression machine. The load at which failure occurs is recorded as the crushing strength of the rock. The completion of the test is usually quite evident since it normally occurs explosively. The results of the test are reported in terms of crushing strength in tons per square foot.

7.2.2 Rock quality. An important factor in determining bearing and excavation conditions of rock is the layered and fractured condition of the rock as revealed by the cores. Due to the spinning of the core barrel in the rock formation and the wash water or air used to carry away the cuttings from the bit, soft zones within the rock will be lost, and the amount of rock recovered in the core barrel often does not equal the length of the coring run. Core runs range typically from 5 to 10 feet depending on the length of the core barrel used. Two numbers are determined to assess the "intactness" of the rock formation: the Core Recovery percentage and the Rock Quality Designation percentage. The Core Recovery percentage is obtained simply by noting the amount of core recovered in the core barrel compared to the length of the core run. For example, if the core run was 60 inches and the amount of core recovered was 48 inches the Core Recovery percentage would be 80.

The Rock Quality Designation is determined by excluding all pieces of the core that are less than 4 inches in length and then repeating the procedure as with the core recovery process. These values are usually reported on the boring logs in the vertical scale of the core runs. These measurements can be done in the field or in the laboratory. When recorded on the boring logs, the designations are usually abbreviated using CR for core recovery and RQD for rock quality designation.

7.3 Laboratory Testing of Soil

7.3.1 Index properties. Index-property tests are generally faster and less expensive to perform than engineering-property tests. These include tests on undisturbed and disturbed soil samples, such as the moisture content of the soil, grain size information, and plasticity. This information is useful for description, classification, and indication of various engineering properties. The unit weight of undisturbed samples can also be determined as an index property. Index-property tests, because of their less-expensive nature, can be done in greater numbers than engineering-properties tests; this permits correlation of strata between borings and extends the engineering properties test results. Much information concerning probable engineering-properties can be estimated from the index-property tests, and a number of correlations have been made in the engineering literature.

7.3.1.1 Moisture content. The moisture content of a soil sample can be determined quite readily in the laboratory. The procedure is to take the sample in the condition in which it was obtained in the field, weigh a small amount of it, dry it in an oven at a specified temperature for a given period of time, and weigh the sample again. The difference in weight is the loss of water. The water lost is divided by the remaining dry weight of the soil specimen to obtain a percentage of moisture content. Typical soils can have moisture contents ranging from 5 percent to 60 percent. Unusual soils may have a much wider range than this. Obviously dune sand sampled from desert locations can be quite dry, while Mexico City clays frequently have water contents of well over 100 percent. This type of water-content test is termed gravimetric water content since it is based on weight. Moisture-content tests should be performed for all lab samples since they are relatively simple to do and can provide useful information. For example, in a soil profile containing sand, silt, and clay zones, the natural water contents will be higher in the clays, due to the greater water holding capacity of clay. This can aid in locating the changes in the formation. Test results frequently are plotted as a profile of water content versus depth on the finished boring logs in a geotechnical report. Figure 6.13 shows moisture content versus depth plotted on a boring log.

7.3.1.2 Grain size. Grain sizes of soil materials can range from boulders and cobbles down through gravel, sand, silt, and clays. A sieve analysis is used to determine the amount of the various sizes present. These can be plotted to provide a "grain-size distribution curve," which in turn can aid in classification and estimating certain engineering properties. For classification purposes sometimes only certain sizes are tested because the break-over in the Unified Soil Classification System between coarse grain soils and fine grain soils is at the #200 sieve size. Grain-size tests are typically done by determining the dry weight of a sample, either by oven drying in advance or by subtracting the moisture content from the total weight, which is the preferred procedure for fine-grain soils. Most grain size distributions of soil materials are done by wet sieving; that is, the known amount of soil is washed through a nest of sieves, which become progressively smaller down to the #200 size at the bottom. After considerable washing and agitation, the sieves are separated and the portions retained on each size are determined after drying. The relative weights retained on each sieve size are divided by the total dry weight of the sample to provide a percentage passing each sieve or retained on each sieve, depending on local practice. Figure 7.1 shows various sieve sizes used to obtain the grain size distribution curve.

The wet sieving procedure typically stops at the #200 sieve size; however, the grain size distribution curve can be extended beyond the #200 sieve by using a hydrometer test in which the smaller sizes of the sample (actually from the #40 sieve on down) are mixed with water, stirred vigorously, and the mixture is permitted to settle in a cylinder under quiet conditions. Periodically a hydrometer is inserted into the liquid, and the depth to which it settles into the water is read at timed intervals. The more soil that remains in suspension the more dense the liquid, and the hydrometer will float higher. As time goes on the soil particles settle and the hydrometer will float lower. This is a fairly complicated procedure, and variables such as temperature must be controlled. This procedure can be used to separate the silts from the clay sizes. Clay sizes are those smaller than 0.002 mm. Since individual grain sizes cannot be seen by the naked eye after they pass the #200 sieve, you can imagine how small 0.002mm is. The #200 sieve actually measures 0.074 mm. Figure 7.2 shows a grain size distribution curve, and Figure 7.3 shows a hydrometer test.

FIGURE 7.1
VARIOUS SIEVES FOR SEPARATING GRAIN SIZES

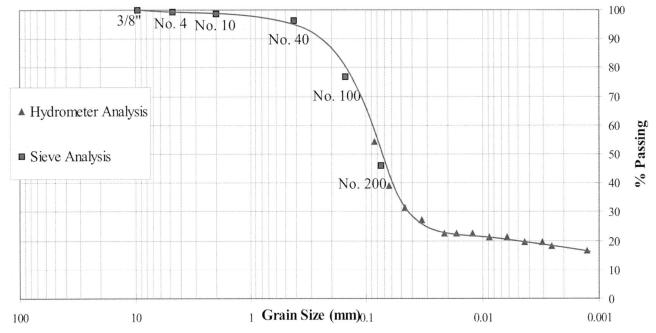

FIGURE 7.2
GRAIN-SIZE DISTRIBUTION CURVE

FIGURE 7.3
HYDROMETER TEST IN PROGRESS

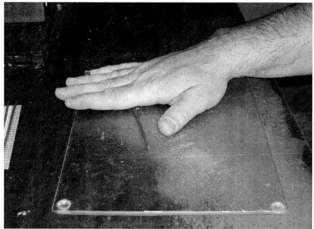
FIGURE 7.4
LIQUID LIMIT AND PLASTIC LIMIT TESTS

7.3.1.3 Plasticity. Plasticity refers to the property of soil remaining in a plastic state over a range of water contents. "Plastic" means that it will hold together while being deformed, as opposed to crumbling and breaking apart. The most commonly used test for plasticity is the "Atterberg Limits," which are named for a Swedish soil scientist who developed this procedure early in the 1900s for agriculture purposes. The process has been adapted by A. Casagrande for engineering purposes and is used extensively in the Unified Soil Classification System.

There are three primary numbers obtained from the Atterberg Limits testing. The first is the liquid limit, which is the water content of a soil at which it changes from a plastic state to a liquid state based on an arbitrarily defined test. The test consists of mixing the material at approximately the liquid limit and placing it in a brass cup. A groove of a defined dimension is cut into the soil in the cup, which is then dropped one centimeter onto a specified impact block. This is done with a rotating-cam device. The liquid limit is defined as the water content at which the groove just flows together on the 25th drop.

The next number that must be determined is called the plastic limit, which is the point at which the moisture content of the soil is such that it transitions, through reduced water content, from a plastic state to a nearly solid state and begins to behave like a brittle material. That is, it cracks up when deformed. This test consists of mixing the soil to a much drier state than the liquid limit condition and rolling to obtain $\frac{1}{8}$-inch threads, usually on a glass plate. When the threads begin to break up at about $\frac{1}{4}$-inch intervals and the thread is $\frac{1}{8}$-inch in diameter they are collected and the moisture content determined. It is an interesting sight to see grown people rolling these little threads of soil out on a glass plate until things are just right. The process can become quite boring, and an occasionally seen error is for the technicians to stop rolling (drying) too soon and reporting a plastic limit higher than it should be. Figure 7.4 shows the elements of the Liquid Limit and Plastic Limit Tests.

The third number obtained from Atterberg Limits is called the plasticity index, which is simply the difference obtained by subtracting the plastic-limit moisture content from the liquid-limit moisture content. This is also known as the "PI." Figure 7.5 illustrates an example of the relation of the water content of a soil to the liquid limit, plastic limit, and plasticity index. Different soils will have different values of the several Atterberg Limits. It has been found that expansive clays exhibit more shrinking and swelling activity the higher the PI. PI is also a useful indicator of a compressible soil if the soil has a high water content; this would differ from an expansive soil that has a low water content in the field but may have similar Atterberg Limits.

A fourth Atterberg Limit sometimes used is the shrinkage limit, which is the water content of a soil at which no further volumetric shrinkage will occur with reduction of water content. This is a smaller value than the plastic limit in clays.

7.3.1.4 Soil suction. Soil suction is a relatively new concept used by geotechnical engineers for analysis of unsaturated soils. The majority of the near-surface soils around the world are unsaturated, and the classical concept of saturated soil mechanics does not completely apply. Unsaturated soils are those in which the void spaces between the mineral grains are not completely filled with water. This concept is especially useful for analyzing and predicting movement in expansive clays.

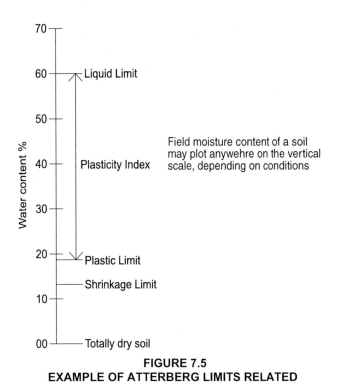

FIGURE 7.5
EXAMPLE OF ATTERBERG LIMITS RELATED TO WATER CONTENTS

Soil suction is expressed in several ways; the most common is using the term "pF." The pF is the logarithm of the height of a column of water in centimeters that can be supported by the negative pressure of the soil because it has a thirst for water. Soil suction in nature typically ranges from about 4.5pF on the dry side to about 2.5pF on the wet side. For example, a pF of 4.0 is equivalent to a column of water 10,000 cm tall being supported by negative or suction pressure in the soil. For perspective, this is over 300 feet. It is true that free surface water, in which typical suction pumps cannot lift water to heights greater than somewhere between 16 and 30 feet based on atmospheric pressure, cannot support itself in a negative state to 300 feet. At this lift in a suction lift pump the water will cavitate. However, in the microscopic structure of a typical soil much higher negative pressures can be achieved and can be measured.

Figure 7.6 illustrates the typical relationships of soil suction to other soil properties.

Soil suction is measured in the laboratory using a dry filter paper method in which the filter paper is permitted to absorb the moisture of the sample in a carefully sealed and temperature-controlled container, and the difference in weight of the filter paper is measured by a very precise scale. Various brands of filter paper have been calibrated to relate weight change to suction levels. Other methods include the use of psychrometers or other devices to more directly measure suction.

The Post-Tensioning Institute procedures for designing shallow soil-supported foundations on expansive clays makes extensive use of the soil-suction concept. Many laboratories are not equipped to do soil suction testing, but the number is increasing, and the use of this term will become increasingly prevalent in the geotechnical world.

7.3.1.5 Expansion Index. The expansion index test is referenced in Section 1803.5.3 of the 2012 IBC and is used in many parts of the United States. The reported index is used for assessing the swelling properties of soil for foundations. The procedure consists of placing a remolded sample in a container that is constrained by a spring-loaded plate. When water is added to this material it swells and moves the plate against the spring, producing a reading on a dial gauge. The initial or starting moisture condition of the sample is important, and this is not clearly defined (40 percent to 60 percent saturation) but depends on local practice and the judgment of the geotechnical engineer. Table 7.1 indicates the level of expansiveness indicated by various expansion index readings.

TABLE 7.1
CLASSIFICATION OF EXPANSIVE SOIL

EXPANSION INDEX	POTENTIAL EXPANSION
0-20	Very Low
21-50	Low
51-90	Medium
91-130	High
Above 130	Very High

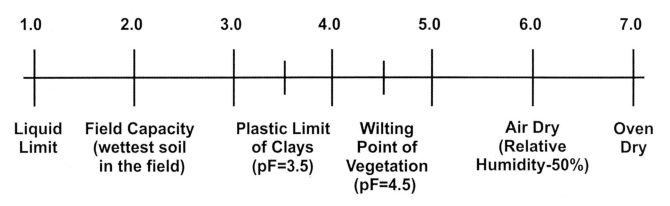

FIGURE 7.6
SOIL SUCTION IN pF APPROXIMATELY RELATED TO VARIOUS SOIL CONDITIONS

The detailed procedure for determining the Expansion Index is set forth in ASTM Standard D4829.

7.3.2 Engineering properties. Engineering properties such as strength, compressibility, permeability, and expansiveness are the values that are actually desired as a result of laboratory testing. However, engineering property tests require elaborate equipment and careful attention to detail by highly trained personnel. Therefore, they take more time and are much more expensive than the index property tests described above. The index property tests are useful in estimating engineering properties and also can be used to extend the results of the various engineering properties tests throughout the site. In other words, soils with similar index properties likely have similar engineering properties in a local setting.

7.3.2.1 Strength. Strength is tested in the laboratory, typically by the unconfined compression test procedure or sometimes by the triaxial compression procedure. The unconfined compression test consists of taking a cylindrical undisturbed sample, usually obtained with the thin-wall Shelby sampler in the field, trimming it square, and pressing it in a machine. The load and the amount that the cylinder of soil compresses at various loads is recorded. The point at which the sample loses the ability to carry the increasing load will be the unconfined compressive strength. Compressive strength samples should have a height approximately twice the diameter to avoid end-cap interference with the strength results. The test is performed at a relatively slow rate of loading, with the loading being performed over about a 2- to 5-minute period of time. Results are generally reported as unconfined compressive strength, sometimes abbreviated as q_u, and reported in terms of tons per square foot (tsf) or pounds per square foot (psf).

The triaxial compression test is similar to the unconfined compression test in that the sample preparation is about the same. However, in the triaxial test a thin rubber membrane is placed around the specimen and lateral confining pressure is applied, either by the use of water or air. There are various kinds of triaxial compression tests depending on what is to be determined. The "Q" test involves simply loading the specimen in a similar way as the unconfined compression test, however with confining pressure added. The same information is derived and reported. Confinement typically is set at the approximate pressures that the overburden in the field would produce on the soil, or confinement can be as directed by the geotechnical engineer. This test is better than the unconfined compressive strength test at determining the actual matrix strength of a soil if it has joints and fractures.

Other variations of the triaxial test include permitting the sample to consolidate under a given confining pressure and then testing it without allowing any of the remaining internal water in the sample to exit the testing device. This is called a consolidated undrained test, also known as an "R" test. Another similar test is called the consolidated drained test, also known as the "S" test. These last two tests are not frequently done because they may require as long as two weeks to complete. However, certain projects require this information.

Usually any of the laboratory strength tests are also accompanied by supporting tests using the Pocket Penetrometer or a Tor-vane device. Other information that should be reported includes the index properties of the material and its unit weight. Figure 7.7 is a photo of a triaxial or confined compression test ("Q" Test) in progress.

7.3.2.2 Compressibility (settlement). The engineering property of compressibility is important in evaluating possible settlement of a structure. There are three major causes of construction settlement: consolidation of fine grain soils by water being squeezed out of them, also known as "hydrodynamic settlement"; settlement of coarse-grain natural soil deposits that are in a loose state; and settlement of man-made fills. There are other serious but less common causes of a structure "going down," which is another term for settlement, such as drying shrinkage, collapsing soils, or area subsidence. Natural deposits of sand or fine gravel are routinely tested by the Standard Penetration Test during field exploration.

FIGURE 7.7
TRIAXIAL COMPRESSION TEST DEVICE

Hydrodynamic settlement or consolidation of fine-grain soils is a result of water being squeezed out of the soil by increased pressure from the new construction. Estimates of the amount and time rate of such settlements can be made based on the laboratory consolidometer test. Figure 7.8 shows a laboratory consolidometer device used for testing the compressibility of a soil specimen.

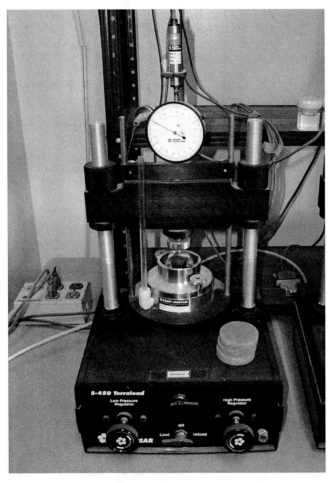

FIGURE 7.8
CONSOLIDOMETER TEST DEVICE

In this test a 2.5-inch diameter by $^1/_2$-inch-thick undisturbed soil specimen is carefully trimmed into a testing ring with porous stones top and bottom. Load increments are applied, and the amount of consolidation or squeezing that occurs is measured with dial gauges. Typically, loads are applied by doubling each load every 24 hours with the initial loads increased by more than doubling. Normally 6 to 9 days is sufficient time for the test. Load ranges are selected by the geotechnical engineer, but generally start very low, such as 0.1 tons per square foot and range up to 32 tons per square foot. The object is to develop a full compression curve of the soil for various loads. From this curve and numerical data derived from the test, engineers can estimate the total amount of settlement that might take place under the applied loads. The plotted movement versus time obtained during each 24-hour loading period are known as "time curves," which can be used to estimate the time rate of settlement.

Since many natural soil deposits have undergone some past consolidation, either due to near-surface drying or from previous loads that have been removed by erosion, it is important for the engineer to know whether the loads to be applied will put the soil mass into the "primary" settlement curve. If the loads are not great enough to reach this breakover point on the compression curve, the loads are within the "pre-consolidated" range. If the imposed load does not exceed the pre-consolidation load, only a small amount of settlement will take place; see Figure 7.9.

7.3.2.3 Expansion. The engineering property of expansion (swelling) can be estimated with the help of the index properties discussed under 7.3.1 and correlations that have been derived using these properties. For example, the plasticity index (PI) of the soil has been found to correlate reasonably well with a potential for swell of a soil. The higher the PI, the greater the expected swell if the water content is low. Table 7.2 depicts one classification of swelling soil versus PI.

TABLE 7.2
CLASSIFICATION OF SWELLING SOIL
BASED ON PLASTICITY INDEX (PI)

PI	SWELL CLASSIFICATION
0-15	Very Low
15-20	Low
20-30	Moderate
30-40	High
Over 40	Very High

To measure directly the amount of expansion that might take place in a soil under field conditions, an undisturbed sample is placed in a consolidometer such as described in 7.3.2.2. This procedure is more accurate than the PI estimating method described above, but is considerably more time consuming and expensive to perform. Since the soil is in a relatively dry state, confining the soil in the consolidometer and adding water will cause the clay to attempt to expand. A full curve of expansion can be obtained similar to that for compression by simply adding water and noting the amount of expansion under given confining loads. A "free swell" test simply measures the amount of expansion of the specimen height that takes place under virtually zero confining pressure. Swell pressures can be obtained for various amounts of expansion. One useful indicator is to obtain the zero swell pressure, which is often measured in terms of several tons per square foot. If the expansive soil is restrained from swelling it can exert tremendous pressure. However, if it is partially restrained from swelling by application of reduced pressures, the expansion will go up, with the zero-restraint or free-swell condition producing the maximum amount of swell. Figure 7.10 illustrates the relation of confining vertical pressure to the percentage of swell of a typical swelling clay.

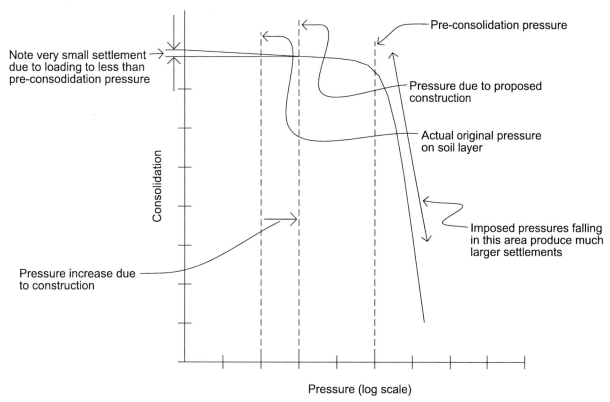

FIGURE 7.9
SOIL CONSOLIDATION CURVE SHOWING EFFECT OF PRE-CONSOLIDATION PRESSURE ON SETTLEMENT

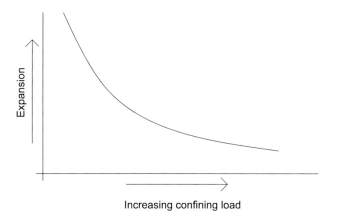

FIGURE 7.10
**RELATION OF CONFINING VERTICAL
PRESSURE TO EXPANSION**

Other methods of estimating expansion include the determination of the "suction-compression index." This value is similar to but not the same as the compression index obtained from the consolidometer test. The suction-compression index is generally called "γ_h" and is a measure of the amount of volume change that will occur in a clay soil with a change in suction levels. By estimating the probable change in suction levels beneath a structure, the engineer can apply the γ_h along with other factors to determine the probable amount of swell.

γ_h can be measured directly by a very elaborate procedure called the COLE test, which determines the change in volume versus a change in suction. Very few laboratories are equipped to handle this test, and a procedure is available to determine γ_h based on index properties, such as the Atterberg Limits, the percentage passing the #200 sieve, and the percentage of clay sizes (0.002 mm). The procedure is presented in some detail in the Post-Tensioning Institute (PTI) Standards entitled *Standard Requirements for Analysis of Shallow Concrete Foundations on Expansive Soils, Third Edition.* See Chapter 35 of the 2012 IBC. The PTI procedure is referenced directly in Section 1808.6.2 of the 2012 IBC and in a number of the codes preceding it as a method for designing shallow soil-supported slabs-on-ground for expansive clay soils.

7.3.2.4 Permeability. Another basic engineering property of soils is permeability. Permeability is a measure of the rate of water flow through soil and is used for designing earth dams and pond liners and estimating flow for dewatering soil. Permeability is measured in the laboratory by use of a permeameter, which is a cylinder containing the soil and a method for measuring water flow through it. Measurements are readily done for sands and silts; however, clay soils are more difficult to measure. This is because clays have very slow permeability and the test is of long duration with errors due to lack of saturation, evaporation, and swelling of the soil.

Permeability can also be measured in the field by a pumped draw-down test in a bore hole with monitor water level wells around the pumped hole. The details of this method as well as the lab procedures are beyond the scope of this book.

The permeability value is designated "k" and can be reported as centimeters per second or inches per minute. Flow can be computed using the k value, the water head differential, and a cross sectional area. The range of permeability can permit water movement that can vary from 1,000,000 feet per year for a coarse sand or gravel to about an inch per year for a high plasticity intact clay.

7.3.3. Other tests.

7.3.3.1 Sulphates. The presence of sulphates is mentioned in Section 1904.1 of the 2012 IBC, which refers to Section 4.3 of ACI 318-11. Sulphates in soil in concentrations above 0.10 percent possibly produce damaging effects on concrete. If the sulphates found in the soil exceed the threshold amount, the building codes require different types of cement and different water-cement ratios in the concrete. Sulphates present in the soil can be tested by wet chemistry, generally by a specialized laboratory.

7.3.3.2 Acidity. It is often of interest to the designer to know the acidity of the soil (or its pH) since highly acid soil can attack and corrode metals, such as a buried metal pipeline or exposed steel piling. Acidity can be tested in the laboratory by the use of a pH meter. The pH scale runs from 0 to 14 with 7.0 being neutral. Readings below 7.0 are acidic and readings above 7.0 are called basic (or alkaline). The further away from 7.0 readings in either direction the more intense the concentration is of being either acidic or basic. The level of acidity in the soil could also influence the effectiveness of lime stabilization, which can be used in pavement or foundation construction.

7.3.3.3 Organics. If organic content is suspected to be significant in a soil profile, tests can be performed that involve the liquid limit test of the original soil compared to the liquid limit after oven-drying at 220 degrees F. If the liquid limit of the oven-dried soil is reduced to 75 percent of the original, the soil is considered organic. Organic contents of soils can range from zero to nearly 100 percent, such as peat. Organic soils with percentages over 5 percent tend to be highly compressible and may not be suitable for supporting foundations of any sort.

7.4 Low-tech or Expedient Estimates

Many of the laboratory tests described above can be approximated in the field by a trained technician to estimate the values that would be obtained by laboratory testing. This is frequently done to provide the soil classification. From these estimates certain relationships to engineering properties can be made, such as strength, compressibility, or expansion.

7.4.1 Shake test for silts and sands. If a fine-grain soil is suspected of having a significant amount of silt or sand, which would change its classification, a procedure called the "shake" test may be used. In this procedure a pad of soil that will fit in the palm of a hand is mixed with sufficient water to become a viscous fluid at about the Liquid Limit. The technician then bangs the side of her or his hand against a table or the other hand, causing a vibration. Water will visibly come to the surface of the soil pat forming a sheen if there is significant silt or sand in it because the permeability of the silty or sandy soil is high and water can easily move through it. This is also known as "water mobility." Soils with high clay content will not have this property; water will just simply stay bound to it and not be able to move through it rapidly. See Figure 7.11.

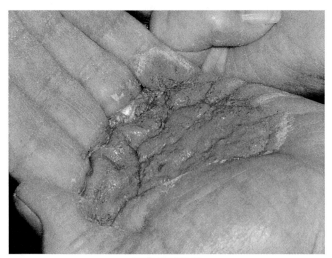

FIGURE 7.11
WET SHAKING TEST

7.4.2 Ball and ribbon test for plasticity. To determine if a clay soil is very silty or is a high-plasticity clay, this test may be performed. The procedure is to obtain a sample of material mixed with sufficient water to become barely plastic, which would be slightly wetter than the Plastic Limit, roll it into a ball and then extrude the ball between the thumb and forefinger to form a ribbon. A long ribbon indicates a high plasticity soil or a high PI. A short ribbon that breaks readily indicates a siltier soil, and the classification may be CL or a combination of CL (clay) and ML (silt). The ribbon test was derived from traditional bakers' practice in that dough was checked to see if it would ribbon long. If it ribboned long, more shortening had to be added until the dough was just right (ribboned short) to produce flakey pie crusts. Figure 7.12 illustrates the ribbon test.

Another ball of soil can be rolled up and set out to air dry. If it is high-plasticity clay it will dry rock hard and be almost impossible to break with the fingers. If it is low-plasticity clay or very silty clay it will not have much strength even when dry.

7.4.3 Strength estimate by fingers, Pocket Penetrometer, or Torvane. Strength of the soil from samples can be estimated in the field by squeezing with the fingers, using a Pocket Penetrometer, or using a Torvane. These are very simple and quick tests to perform, but will give a reasonable estimate of the strength of the soil. Table 7.3 lists the strength estimated by indentation that can be made with fingers.

TABLE 7.3
SOIL CONSISTENCY

CONDITION	UNCONFINED COMPRESSION (TSF)	FIELD ESTIMATE
Very Soft	<0.25	Squeeze out between fingers
Soft	.25-.50	Molded by light finger pressure
Med. Stiff	.50-1.0	Molded by strong finger pressure
Stiff	1.0-2.0	Can be indented by thumb
Very Stiff	2.0-4.0	Can be indented by thumbnail
Hard	>4.0	Difficult to indent by thumbnail

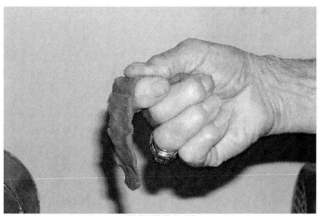

FIGURE 7.12
RIBBON TEST

The Pocket Penetrometer is a device that is pressed into an intact piece of the soil until it penetrates $1/4$ inch. Then a spring-loaded calibration is read, which indicates the equivalent unconfined compressive strength in tons per square foot. A Torvane is a device in which small sharpened vanes set in a circle typically 1 inch in diameter is pressed into an undisturbed soil sample and twisted. The amount of torque needed to move the vanes by shearing the soil is recorded and translated into shear strength in either tons or pounds per square foot. The Torvane is best used to test softer soils. See Figure 7.13, which is a photograph of a Pocket Penetrometer and a Torvane.

7.4.4 Estimate of grain sizes by visual methods.

Grain sizes can be determined in a soil sample by a visual examination depending on the nature of the soil. If it is all a fine-grain soil, the procedure is straightforward, and everything appears to be smaller than the #200 sieve. A good rule of thumb is that individual soil particles cannot be seen with the naked eye if they are smaller than the #200 sieve. If a sample of this type of material is thinly smeared in the palm of a hand and individual particles cannot be detected, it is a likely indication that the material all passes the #200 sieve. If a soil sample has clay or silt mixed with sands, one procedure is for the technician to take a small bite of the soil between his or her front teeth. If there is a gritty feel, there is sand in it (larger than the #200 sieve), if not, it is all clay or silt. Not everyone enjoys doing this test. Another procedure is to take a sample believed to be a mixture of sand and clay and place it in a small glass container, such as a fruit jar, mix it thoroughly with excess water, shake it

up, and set it on a still surface. The material that settles to the bottom within the first five seconds will be larger than the #200 sieve. A person can make an estimate of the percentage of sand by knowing the volume of the original sample before the material larger than #200 sieve sizes was washed out and settled.

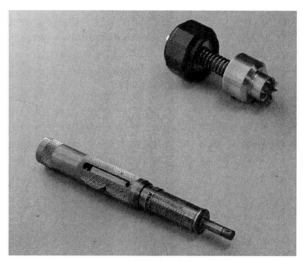

FIGURE 7.13
POCKET PENETROMETER AND TORVANE

For coarse-grain materials without much fine grained soil, the best procedure is to spread the sample out on a table or canvas in the field and hand pick the larger pieces and put them on one end, put smaller pieces in the middle, and put the very small sizes on the other end. By separating the materials this way a rough visual estimate of the percentages of each size can be obtained. Figure 7.14 is an illustration of the visual examination of coarse-grain soils.

7.4.5 Estimate of expansion potential.

A field estimate of expansion potential would involve noting whether the soil material has medium to hard consistency and if it is CH clay. If it is CH clay and the moisture content appears to be close to the plastic limit, as evidenced by its relationship to the ribbon test, it is likely to have potential for expansion. It is not possible to quantify potential expansion by this test, but it will give an indication that expansion is a significant problem that should be examined further. If a stiff to hard CH soil is jointed and fractured, it is likely that it is an expansive clay.

FIGURE 7.14
VISUAL EXAMINATION OF COARSE-GRAIN SOILS

7.4.6 Field description and classification.
When doing a field exploration followed by laboratory testing and sample examination, the engineer or technician should describe the material encountered at each depth and note its probable classification on the log. A good rule is to indicate the kind, condition and arrangement of the soil.

The kind of soil material should be noted on the log as to the depth at which the material begins and the depth at which it ends. The description of the kind of soil includes its basic descriptor such as clay, sand, gravel, etc. The description should also include modifying elements such as silty clay, clayey sand, or gravel with clay. Additional information should include the color, the maximum grain size observed, the estimated percentage of sand and gravel, and the sample's USCS classification.

The soil's condition would include its moisture content and estimates of its strength. Strength estimates for fine-grained soil could include the Pocket Penetrometer or Torvane results, or consist simply of an estimate from finger procedures. If conglomerate (a cemented mixture of sand, gravel, and fine-grain soil) is found, the degree of cementation should be noted.

The soil's arrangement refers to the density condition of the soil: either loose or dense. This refers to coarse-grain materials such as sands, and this estimate is primarily determined by the difficulty of drilling and the blow count or "N" values obtained from the Standard Penetration tests. Coarse materials should also be described as "well" or "poorly" graded. Layering, joints, or fractures in the soil or rock are included in "arrangement."

Figure 6.13 in Chapter 6 illustrates a hypothetical boring log. The discussion of kind, condition, and arrangement is included by various methodologies. Some laboratory properties can be displayed on the boring log for clarity and ease of interpretation, such as the profile of moisture content, strength estimates, Atterberg Limits, and laboratory test results concerning key grain size percentages.

7.5 TEST QUESTIONS

MULTIPLE CHOICE

1. In determining the strength of soil in the laboratory, what is meant by a "Q" test? (select one):
   a. Unconfined Compression Test
   b. Confined Compression Test
   c. Confined Compression Test with drainage

2. What are the three major causes of construction settlement? (select three):
   a. hydrodynamic settlement
   b. poor inspection
   c. equipment breakdown
   d. coarse-grained soils in a loose state
   e. settlement of man-made fills
   f. a natural deposit containing boulders

3. Index property tests (select two):
   a. are generally faster and less expensive to perform than engineering property tests
   b. can be confusing because of the large number of tests
   c. can extend engineering property test results to other parts of the soil profile
   d. may be used to replace engineering property tests

4. Grain sizes of soil materials can range from (select one):
   a. gravel to silt size
   b. boulders down to silts and clays
   c. sand to clay sizes
   d. boulders to fine sand size

5. In a natural soil deposit, loads being applied to the soil mass in the primary settlement range (select one):
   a. will cause settlement less than those in the pre-consolidated range
   b. will cause settlement more than those in the pre-consolidated range
   c. will always result in shear failure

6. Organic soils will (select one):
   a. most likely be suitable for establishing foundations
   b. most likely will not be suitable for establishing foundations
   c. cause extensive decay of rock

7. If a stiff to hard CH soil is jointed and fractured (select one):
   a. it is an indication of heaving loads from a previous foundation
   b. it is likely that it is expansive clay
   c. it cannot support footings

8. If a core run was 120 inches and the amount of core recovered was 100 inches, the core recovery percentage would be (select one):
   a. 100 percent
   b. 12 percent
   c. 83 percent
   d. 60 percent

9. The break-over in the Unified Soil Classification System between coarse-grain soils and fine-grain soils is at the (select one):
   a. #100 sieve size
   b. #4 sieve size
   c. #10 sieve size
   d. #200 sieve size

10. A generally accepted soil-suction condition defined as the wilting point of vegetation is pF= (select one):
    a. 2.5
    b. 3.8
    c. 4.5
    d. 6.0

11. In terms of soil suction, a pF of 4.0 is equivalent to the negative pressure of a column of water (select one):
    a. 400 centimeters high
    b. 40 centimeters high
    c. 1000 centimeters high
    d. 10,000 centimeters high

12. An Expansion Index indicating a medium potential expansion can be a range of (select one):
    a. 21 to 50
    b. 91 to 130
    c. above 130
    d. 51 to 90

13. Soil expansion can be estimated using the Plasticity Index on a low-water content clay soil. Some relationships of swelling indicate that a Plasticity Index of over 40 is (select one):
    a. high
    b. moderate
    c. very high
    d. low

14. A standard published by the Post Tensioning Institute entitled *Standard Requirements for Analysis of Shallow Foundations on Expansive Clay* is referenced directly in 2012 IBC Section (select one):
    a. 1609.1.1
    b. 1808.6.2
    c. 1808.6.4

15. In the strength estimate of soils by field-expedient methods using finger procedures, a strength equivalent to an unconfined compression strength of 1.0 to 2.0 TSF can be estimated by the soil being able to be indented by (select one):
    a. thumbnail pressure
    b. difficult to indent by thumbnail
    c. strong finger pressure
    d. thumb pressure

Chapter 8

ANALYSIS OF SITE INFORMATION AND CONSTRUCTION DOCUMENTS

Site investigation reports

Chapter 8

ANALYSIS OF SITE INFORMATION

8.1 Soils Report Contents

The soils report (geotechnical report) should contain the following:

- Identification of the client.

- Authorization to do the work and issues to be investigated.

- Description of what investigations and tests were done.

- Results of field and laboratory investigations including the test data, boring logs, and boring location plan.

- Discussion of the general site conditions including topography, geology, and soil conditions.

- Analysis and discussion.

- Conclusions.

- Recommendations.

Recommendations should include the engineer's specific recommendations as to the type or types of foundations that might be suitable for the site and the design values that should be used. Design values may include the allowable bearing pressure, depth of footings, skin-friction values, lateral-load forces, and lateral resistance that may be available for the various parts of the foundations. There should also be information concerning constructability and quality control related to earthwork and foundations and, in many cases, special specifications, such as information relating to select fill or stabilization procedures. See Section 1803.6 in the 2012 IBC.

8.2 Determining Type of Foundation

There are basically two types of foundations—shallow and deep. Shallow foundations can include spread footings, strip footings, or a shallow raft or mat. The bearing portion of footing walls should be designed as a shallow strip footing. Deep foundations can include piling, piers, or deep-excavated basement footings. Sometimes combinations are used, such as deep foundations to support the main frame of a structure with a soil-supported slab on the first floor.

In the selection of the type of foundation, the engineer should consider the function of the structure, loadings, and the sensitivity of the structure to movements. For example, typically a warehouse can tolerate more differential movement and slight distress than could a monumental building or a custom home. If there is low-bearing or high-settlement type soil from the ground surface down to a certain

depth over a stronger soil, it may be valid to use deep foundations to penetrate through the weaker material, transferring the load to the stronger material. The same is true for non-engineered fills, which may not be reliable to support near-surface construction. Shallow foundations may be appropriate if sufficient bearing pressures are available near the ground surface and the weight of the structure will not generate unacceptable settlements. In many parts of the country, residential and light commercial structures are almost universally built on shallow foundations that typically are stiffened slabs-on-ground or non-stiffened uniform thickness slabs-on-ground. "Stiffened" refers to a pattern of ribs or beams underneath the foundation slab that provide additional bending stiffness.

Occasionally shallow foundations are not suitable because of high shrink-swell activity of the near surface soils, and a deep foundation may be more appropriate unless procedures to reduce the shrink-swell differential near the surface are employed.

8.3 Allowable Bearing Values

Allowable bearing values are most critical in medium to soft soils, and these types of soils require careful calculation. A typical method for calculating these values in clay would be to determine the unconfined compressive strength (q_u). After going through several calculations, such as bearing capacity factors and factors of safety, a safe-bearing value in most footings of shallow to medium depth will be found close to the actual unconfined compressive strength (q_u). For deeper footings, such as piers, the end-bearing pressure is somewhat higher, on the order of 1.5 times the unconfined compressive strength. A quick check of bearing can be made in the field using the Pocket Penetrometer.

If the material is granular and has a property called internal friction, the bearing values are calculated differently. This methodology is beyond the scope of this book. For sand deposits, bearing is generally not a problem; however, settlement may be, depending on the density or the looseness of the upper sand layers. This is most readily determined by the split-spoon drop-hammer field test, and charts are available to relate settlement to applied bearing pressure for various N values. Figure 8.1 shows an example of this chart.

Rock can be evaluated for safe-bearing pressures, which of course will be much higher than those of soil. The weakest rock is typically stronger than the strongest soil, and safe-bearing values for rock usually start at about 5 tons per square foot or 10,000 psf. Highly weathered or layered rock may have lower values. Very strong rock can have values that go much higher, up to 80,000 psf or more. The safe-bearing value of rock is determined by crushing cores of the rock obtained during the field investigation as described in 7.2. Table 8.1 illustrates the presumptive values permitted by the *International Building Code* compared to the possible range of values that could be obtained through a geotechnical investigation.

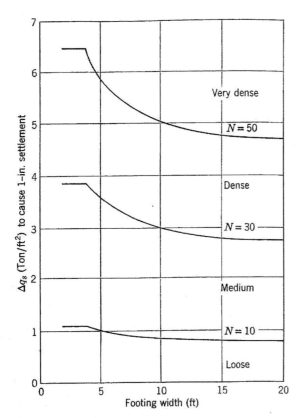

Based on Lambe & Whitman Soil Mechanics © 1969, Figure 14.28 p. 221. Reproduced with permission of John Wiley & Sons, Inc.

FIGURE 8.1
SETTLEMENT OF FOOTING ON SAND FROM STANDARD PENETRATION RESISTANCE

8.4 Settlement Estimates

In an engineering sense, settlement analysis is different from the determination of allowable bearing values although they can be related. On a shallow and local basis, the higher the bearing pressure the more the settlement could be. However, allowable bearing values are established to avoid shearing or tearing the soil immediately beneath the footing, while settlement estimates often have to do with a larger area and deeper layers of soil. Estimates of potential settlement should be done by a geotechnical engineer. In general, if the proposed construction is reasonably heavy, and fine-grain soils exist at a high water content, the potential for settlement is present. The presence of a significant amount of organics is always a cause for concern about settlement. In any event, an estimate of potential settlement using one or more suggested foundation configurations for the proposed building should be given in the report. Potential settlement should be reported as total and differential across the foundation.

TABLE 8.1
COMPARISON OF ROCK SAFE-BEARING VALUES IN PSF

PERMITTED BY 2012 IBC TABLE 1806.2		POSSIBLE VALUES FROM GEOTECHNICAL INVESTIGATION
Crystalline Bedrock	12,000	12,000 to 80,000
Sedimentary and Foliated Rock	4,000	4,000 to 30,000

8.5 Shrink and Swell Estimates

If the engineer identifies expansive clay soils within the upper 15 feet, he or she should always make an estimate of the shrink-swell activity of the soil, both in general terms (such as low, medium, and high potential) as well as numerical values. These problems are usually more significant with lightly loaded structures, such as residences and light commercial buildings; although soil-supported warehouse or shopping center floor slabs may be lightly loaded even though the main building is established on piers. Typical shrink-swell estimates are in the range of $^1/_2$ inch to 8 inches, although higher values are possible. There are various methods to reduce the total amount of swell such as removing clay soils and replacement with inert fill, or injecting water or chemicals, and the engineer may recommend one or more of these to the client.

Numerical values for designing shallow slab foundations should be recommended in the soils report. Uplift skin friction on piers or piles and establishment depths to obtain anchorage against uplift should also be given if deep foundations are an option.

8.6 Skin Friction on Piers or Piles

If a pier or pile is constructed in the ground and the building loads are applied to the top of it, the tendency for the pier or pile to move downward into the earth is resisted both by the tip resistance and by side resistance called "skin friction." Skin friction can be a very important value for use in design for deep foundations and may actually carry the majority of the load of the pier or the pile. The resistance of the soil available to the pier or pile is a variable that depends on the type of soil, the strength of the soil, and the condition of the surface of the pier or pile. Soil can generally be divided into sands and silts/clays. The available skin friction for sand increases with depth due to an increase in the confining pressure as the overburden becomes deeper. The actual values also depend on the internal composition of the sand.

Clay/silt soils have a more uniform skin friction, provided the soil is the same consistency up and down. Soils that are a combination of sands, silts, and clays require more complicated calculations to determine the actual skin resistance. The type of contact surface on the pile or pier is important. For example, a drilled pier in many types of soils will adhere to the soil material when the concrete is placed in the pier because the concrete grout flows into the soil and actually forms a soil-cement layer around the pier, giving a good bond. On the other hand, the action of the auger can smear a softened layer on the surface of the drilled pier hole in a stiff to hard clay or a rock material.

Therefore, the actual skin friction that can be realized in such materials depends on the conditions at the time of drilling, which of course can be improved by cleaning the hole walls. Since cleaning and scraping is typically not done and is difficult to verify in the field, often the allowable skin friction values for these materials is reduced assuming that there will be a softened zone around the pier.

Since a pile is driven to penetrate soil by displacing the soil, causing considerable disturbance, it is likely that the end-bearing and skin-friction values of a driven pile will increase in clay/silt soils after driving has stopped. In effect, the clay is disturbed by driving the pile through it, but with time the extra water that was squeezed out of the immediately surrounding soil will dissipate and the soil will become stronger. This is called soil freezing or soil set-up. A silt/clay soil will nearly always become stronger than at the time of driving and sometimes even stronger than the undisturbed original soil.

Other variations of skin friction include what might be called reverse skin friction or uplift friction; this is caused by expanding clay soils in the upper 5 to 20 feet of the pier or pile. In this case the soil expands and tries to pull the pier or pile out of the ground by transferring uplift forces through skin friction. In such cases the skin friction in the anchorage zone below the expansive zone must be evaluated so the designer is certain that sufficient penetration is achieved to generate hold-down forces to resist the up-lift forces.

In some cases the action of driving piles or even drilling piers in very soft deposits will disturb an open soil structure, and it will begin to consolidate under its own weight, causing down-drag skin friction on the pier or pile. If this is the case, the geotechnical engineer and the foundation designer must be aware of it and allow for these extra forces in their design. See Chapters 12 and 13 for more detailed design factors for piers and piles. The various values discussed above should be reported in the geotechnical report if appropriate to the site.

8.7 Lateral Loads on Walls, Piers, or Piles

In addition to the typically downward gravity loads on the foundation elements, retaining walls, piers, or piles may be subjected to lateral or horizontal loads. These loads are the result of a change in the surface elevation requiring a retaining wall, or these loads are the result of downhill slope instability ranging from creep to the possibility of a large-scale landslip. Seismic forces also typically apply significant lateral loads to foundations.

Retaining walls are used because of a desired change in the ground-surface elevation. Since most soil materials will not stand vertically for very long, if at all, a retaining wall must be supplied to maintain the upper soils in place. Two French engineers, Coulomb and Rankine, in the 1700s and 1800s, derived formulations for determining horizontal earth pressure on retaining walls for different types of soils. The approach has changed little since then, although it has been refined and now can be done on computers. The basic idea reduced to a simple form is to consider the earth behind the wall as a fluid. The horizontal pressure of water

behind a dam increases with depth in a linear fashion. In other words, the water pressure at 20 feet is much higher than the water pressure at 5 feet. Considering the earth behind a retaining wall as "equivalent fluid" permits a similar analysis with values of the equivalent fluid being modified to reflect the properties of the soil. Water weighs 62.4 pounds per cubic foot and has no internal strength, so the full force of the water goes laterally as well as vertically. Soil materials' actual weights range from about 85 to 135 pcf, but because of the internal strength of the soil, the horizontal pressures are typically less than the actual full weight of the soil. Values used for design range from about 30 to over 100 pcf. The highest unit weight has been assigned to soils that may have internal strength but also have the property of swelling and so exert added horizontal force. Table 8.2 is taken from a widely used geotechnical engineering textbook. The table recommends equivalent fluid weights to be used for soils of different descriptions. The values can be used for design of most retaining walls up to 20 feet high, with level soil behind the wall and no additional surcharge, such as vehicle or building loads. Section 1807.2.3 of the 2012 IBC, Section R404.4 of the 2012 IRC, and good practice for retaining walls requires a factor of safety of 1.5 against lateral sliding and overturning.

TABLE 8.2
TYPES OF BACKFILL FOR RETAINING WALLS

TYPES OF BACKFILL
1. Coarse-grained soil without admixture of fine soil particles, very permeable clean sand or gravel
2. Coarse-grained soil of low permeability due to admixture of particles of silt size.
3. Residual soil with stones, fine silty sand, and granular materials with conspicuous clay content.
4. Very soft or soft clay, organic silts, or silty clay.
5. Medium or stiff clay, deposited in chunks and protected in such a way that a negligible amount of water enters the space between the chunks during floods or heavy rains. If this condition cannot be satisfied, the clay should not be used as backfill material. With increasing stiffness of the clay, danger to the wall due to infiltration of water increases rapidly.

1. Based on Terzaghi & Peck *Soil Mechanics in Engineering Practice*, 2nd Ed., ©1967 Table 46.1 Reproduced with permission of John Wiley & Sons, Inc.
2. IBC 2012 Table 1610.1 has a more detailed presentation of values. (See Chapter 22)

The following equivalent fluid unit weights are appropriate for use with the various soil types.

TABLE 8.3

TYPE	EQUIVALENT FLUID UNIT WEIGHT PCF
1	32
2	36
3	46
4	100
5	120

The rigidity of the retaining wall can affect the lateral pressures on it. A very rigid wall, such as a bridge abutment, actually generates higher pressures than one that is slightly flexible, such as a typical cantilever reinforced concrete retaining wall. The very rigid state is called the "at rest" condition, and the slightly flexible state for which the majority of retaining walls are designed is called the "active" state. The geotechnical engineer needs to consider these conditions as well as the soil's properties, and he or she should include suitable recommendations for different types of retaining walls in the report.

Piers or piles may be subjected to horizontal force due to the soil moving laterally, such as from downhill creep. This is a condition that requires careful analysis and consideration, and the loads of the moving soil will probably vary with depth, being higher at the surface and lower as the depth increases. Sometimes the way to analyze this is to assume that the soil actually fails and flows around the piers or piles and is similar to a lateral ultimate bearing capacity failure. This can produce extremely high horizontal forces on these foundation elements. Figure 8.2 illustrates some of the lateral forces that could affect piers or piles.

Seismic shaking produces horizontal loads because the earth itself is moving from side to side while the building, due to inertia, is attempting to stay still. The forces are transmitted to the footings, then to the building, and the building begins to respond. The analysis of seismic forces will be discussed in more detail later, but in seismically active areas this can be the overriding design consideration for foundations.

8.8 Construction Approaches

The geotechnical report should contain suggestions concerning the method of construction. Such considerations could include the need for dewatering excavations, the need for casing piers, and considerations of settling of soil around excavations causing damage to nearby structures. Vibration from pile driving may also cause damage beyond the limits of the construction project and should be discussed in the soils report, if piling is considered. The purpose of the geotechnical report is not to dictate to a contractor how to build a project, but to provide design guidance to the project design professionals and insight for contractors regarding some of the situations that may need to be dealt with in the field during construction.

8.9 Special Specifications

It is frequently appropriate for the geotechnical report to contain guide specifications for certain items such as specialty construction including temporary bracing and shoring, quality control processes relating to the foundations and earthwork, and specifications for select fill. Such specifications are not intended to be directly incorporated in the contract documents, but used as a guide to the person preparing the specifications. See Figure 8.3.

8.10 Aggressive Soils

Aggressive soils can include soils with high sulphate content, acidity, or alkalinity, which could attack foundation concrete or steel or metal utility pipes in the ground. The possible effect on the proposed construction should be discussed in the geotechnical report, and the engineer should give values to quantify the nature of the aggressive soil. Approaches should also be recommended to deal with these problems. Such approaches could include recommending PVC pipe instead of cast iron or steel pipes or wrapping metallic pipes with a protective coating, recommending additional concrete cover over the reinforcing steel for in-ground elements, or using a sulphate-resisting cement in the concrete mix. In some cases a cathodic protection system may be recommended.

8.11 Good Reports/Bad Reports

It may take a specialist to determine if a soils report is adequate or inadequate; however, certain indicators are fairly obvious. They could include lack of a basis for forming opinions, such as no field or laboratory test information; inadequate detail in the boring logs; and disagreement among the data, discussions, and recommendations. For a geotechnical firm that does thousands of geotechnical reports a year, many reports become repetitious, and it is a frequent practice to use previous reports modified for a new report if the conditions are very similar. While there is nothing wrong with this, sometimes errors are introduced, and the report should be read to see if strange references are given not relating to the site in question. The geotechni-

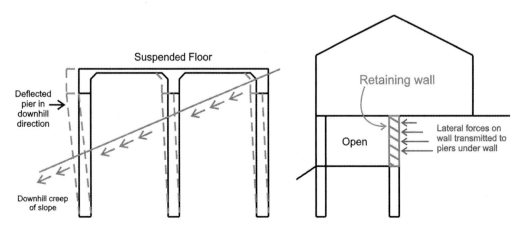

FIGURE 8.2
LATERAL FORCES ON PIERS OR PILES

cal report should consist of a carefully worked-out analysis of valid laboratory and field data, with proper application of seasoned judgment by the engineer. A bad report may be simply a semi-educated guess and contain very little support for the recommendations that are made.

An attempt to use reports that were not done on the site, a long way from the proposed construction, is an indication that a dangerous situation is evolving. It is difficult enough for the engineer to determine the subsurface conditions when he or she has borings placed under the footprint of the building, but if they are a quarter-mile or a mile away the situation becomes hopeless, and such reports should be rejected out of hand. Sometimes a geotechnical study is performed with borings in the field at a certain location, and the project design team moves the building. In this case the situation should be re-evaluated. If the building footprint and the actual boring locations deviate only a small amount, the original borings may still be applicable. However, if there is a large deviation there should be a supplemental report providing additional borings, possibly altering the recommendations. The permissible deviation distances depend on the probable variability of the site and will be a matter of judgment by the geotechnical engineer.

If the report contains laboratory testing results that would permit classification of the soil and yet the classifications reported in the report and in the boring logs are markedly different, it may be an indication that someone was not paying attention. For example, if the material is called a sand in the boring logs and yet it has several Atterberg Limits and sieve tests that indicate that it has a high liquid limit and more than half is passing the #200 sieve, it is probably a clay. Such indicators should be questioned and a determination made if there is a valid reason for this.

8.12 Foundation Construction Without a Formal Geotechnical Report

Many projects are built throughout the United States without a formal geotechnical report. This is acceptable if the structure is small, such as a residence or small commercial building, and there is an adequate amount of history or experience with this type of building in a particular area. An example would be a house foundation in the midst of a subdivision that utilizes a foundation similar to all the houses around it with successfully performing foundations and there are no indications that the site is non-uniform. This is a judgment matter that the structural designer as well as the building official need to consider. If the buildings are of a routine nature and relatively small, the standard code provisions may be sufficient for the design. However, some knowledge of the soil conditions that exist should be known to the designer, such as whether it is shallow rock, deep sand, or stiff clay.

All major buildings or structures such as bridges, dams, and anything that has a life-safety aspect should be designed using a formal geotechnical report. If the structure is to be established in a high seismic risk area, certainly an analysis of the soil would be in order. Likewise, if the site contains fill of an unknown origin, very soft soils, or highly organic soils, a thorough investigation is called for. Expansive clay soils, even for small and light structures, need a geotechnical investigation since the code requires certain types of design procedures to be used for these buildings (Section R403.1.8 of the 2012 IRC and Section 1808.6.2 of the 2012 IBC). The referenced design procedures call for a geotechnical study to support the design.

Select Underslab Fills:

1. Selection of fill material should be guided by the following criteria:
   a. Maximum plasticity index: 17; Minimum plasticity index: 3
   b. Minimum and maximum passing 40-mesh sieve: 10% to 40%
   c. No stones larger than 11/2"

2. Compaction should be to 90% of maximum laboratory density determined in accordance with ASTM D-698, using a compactive effort of 7.16 ft.lbs./cu.in.

3. Placements should be in lifts not exceeding 6 inches after compaction. A 1-inch sand level-up course may be placed immediately below concrete slab and waterproof membrane.

4. Proposed fill material should be submitted to engineer for approval, in advance of placement.

5. Backslopes: Select fill shall extend a minimum of 2 feet, (3 feet for fills over 6 feet high) horizontally before sloping downward at one on two backslope or as per plans to natural soil. Backslope fill shall be well compacted. If perimeter grade beams are carried into natural soil, backslope fill is not required.

6. Interior select underslab fill may be sacked to achieve vertical form of inside of perimeter grade beam. If sacked, density tests will not be taken closer than 3 feet from inside face of perimeter grade beam. No other foundation beams may be formed by sacking.

FIGURE 8.3
EXAMPLE OF SELECT UNDERSLAB FILL SPECIFICATION

For demanding site conditions, design of foundations without a geotechnical study is simply a semi-educated guess. Two bad things can happen from this procedure. The first is that the foundation will be over-designed, and some of the owner's money will be wasted. The second is worse in that a failure could occur in which case all the owner's money will be wasted. Even though a geotechnical study may not always provide the absolute best answers, it is certainly better to have as much information as can be obtained prior to design.

8.13 Good Foundation Plans

8.13.1 Identification. Good plans should identify the name of the project with a specific location, such as an address, and clearly labeled sheets so the scope can be properly presented. The various professional design organizations involved, such as the architect and structural engineer or designer, should be identified in the title block with their addresses and telephone numbers for easy contact. Any plan revisions should be clearly described and dated.

8.13.2 Soil data and design-criteria references. Foundation plans should include the basic soil information, such as allowable bearing capacity or expansive clay soil parameters, and the references for the design procedures and criteria that were actually used to prepare the plan and do the calculations. The particular version of the building code or other design references that may have been used should be listed. A reference to the project's geotechnical investigation report should also be on the plans, generally in the notes section.

8.13.3 Responsibility. A clear line of responsibility should be evident in the plans listing the owner or project sponsor, the architect, the foundation and/or structural designer, and the geotechnical engineer. Engineer-prepared plans should always have the engineer's professional seal and signature clearly displayed, as well as the date of the release of the plans.

8.13.4 Plan features. Foundation plans should have an overall plan view of the foundation layout that references sections and details; this may be on the same sheet or separate sheets as required. The establishment of depth of footing should be clearly shown, and any possible field alternates, such as references to adds or deducts to pier depths, should also be indicated. Since foundation construction is closely related to the earth that supports the foundation, some reference should be made to site preparation such as clearing of top soil; removal of trees and grubbing roots; general preparation of the subgrade, including compaction; and provisions for keeping the site drained and protected from ponding water during and after construction. If below-grade finished areas are involved with the foundation, such as basements or split-levels, details of waterproofing and drainage relief behind the walls should be supplied.

If the general region requires reducing moisture transmission through a slab-on-ground or ventilating areas under a foundation with a crawl space, either details or notes should indicate the methods to be used. Crawl space-type foundations may also require a drainage feature to avoid water accumulating and standing under the foundations as well as methods for underfloor ventilation; see Sections R405 and R408 of the 2012 IRC.

8.13.5 Plan notes and specifications. Foundation plan notes should include the soil design data; geotechnical report references; information on which general commercial specifications are to be employed, such as the American Concrete Institute Specification 318 for concrete and reinforcing steel; the grade of reinforcing steel to be used; and any special notes on splicing reinforcing steel. If structural steel is involved, the AISC specification should be invoked along with the steel-grading specification. A special note may be needed for drilled piers or driven piling work if included in the project, including the need for test pilings or special inspections for these installations. A schedule of construction inspections and frequency should be listed, such as acceptance tests for fill and drainage material, field density, concrete strength tests, pier inspections, and piling inspections.

Sometimes more elaborate specifications are included, such as a specification for a select fill that will be used in certain aspects of the foundation construction. On larger jobs, detailed specifications are given in the "spec book" or project manual.

Figure 8.4 is an illustration of typical foundation plan notes and short specifications. Such notes may need to be modified for each particular project and will vary by locality.

8.13.6 Does the design follow geotechnical recommendations? Good foundation plans will follow the recommendations of the geotechnical engineer. If the foundation designer feels that it is necessary to deviate from these recommendations, he or she should seek a supplemental report from the geotechnical engineer agreeing with the alternate approach to insure that geotechnical considerations were adequately provided for in the design. If the foundation designer decides to be more conservative than the approach recommended by the geotechnical engineer, there is no harm done; however, it is important to insure that the general approach follows that of the geotechnical engineer only made more conservative. For example, if the geotechnical engineer permitted up to 5,000 psf for safe bearing values for a spread footing and the design engineer used 3,000 psf, this would be an acceptable conservative approach. On the other hand, if the design engineer used 7,000 psf, this is a non-conservative approach, and the designer is well advised to have such an approach authorized and documented by a supplemental geotechnical report or letter.

NOTES

1.0 GENERAL

1-1. Inspections by Engineer or his representative required, as applicable, for: fill placement and compaction; tendon placement and forming of beams and slab; concrete placement and testing; tendon stressing; pier inspection.

1-2. Tendon lengths and count and concrete quantity estimate on plan are for estimating purposes only. Contractor should verify all tendon lengths and concrete quantity prior to installation. Concrete quantity must be adjusted for sloping site and forming irregularities. Concrete quantities are not exact. Draped tendons are not shown. U.N.O. for plan clarity.

1-3. Plan shows the location of structural reinforcement, beam depth and beam locations only. Architectural dimensions must be compared to the architectural plans prior to construction of forms. Report any discrepancies to the Engineer. The forms should be built using the architectural plans—not the Engineer's plan. Do not scale plan.

1-4. This design is in accordance with the PTI Design and Construction of Post Tensioned Slabs-on-Ground 2nd Edition, the 1997 Uniform Building, The 2000 and 2003 International Residential Code.

1-5. These plans are copyright MLAW as of the year dated.

1-6. Vertical control joints should be used in exterior masonry to the full height spaced approximately 15 feet apart. A joint should be located directly above all slab control joints.

2.0 SITE PREPARATION

2-1. All site work shall be performed in accordance with FHA Data Sheet 79-G. Refer to notes concerning "approved" and "unapproved" fill.

2-2. All underslab "Forming Fill" shall have a P.I. less than 20 and be free of organics.

3.0 CONCRETE

3-1. Concrete shall have a minimum compressive strength of 3000 psi at 28 days. Concrete should be minimum 2000 psi at full tendon stressing. All concrete work shall meet A.C.I 318. Concrete shall be deposited in forms no later than two hours after water is mixed at the plant. One addition of water will be permitted at the job site to adjust the slump to a maximum of 6 inches.

3-2. Concrete shall be well consolidated using proper mechanical vibration, especially in the vicinity of the tendon anchorage.

3-3. If conduit in slab is required pr or to concrete placement, location to be verified in field. Piping, vents or electrical cables shall be placed so as not to reduce slab thickness. Plumbing and/or conduits larger than 1" diameter must be trenched into underslab fill.

3-4. If unanticipated interruptions in concrete placement occur, and concrete hardens, temporary forms must be used for setting of construction joints or concrete must be chipped to form vertical joints prior to placing additional slab. Use #3 X 24" dowels at 12" O.C. epoxed into existing concrete to bond old to new concrete.

3-5. FLATWORK MAY BE PLACED ONLY AFTER STRESSING.

4.0 CONCRETE COVERAGE

4-1. SLAB TENDONS:
1-1/2 inches above sub-grade in 4" thick slab and ANCHORS to have 4 inches vertical coverage from center of anchor to top of concrete.

4-2. Slab Tendons may be moved 12" max. horizontally to allow for plumbing box-outs. Beam Tendons may be moved 3" downward and/or 2" upward vertically for plumbing/conduit pipes in beams.

4-3. BEAM AND WALL STEEL:
1-1/2 inches slab, 2 inches formed, and 3 inches exposed to earth.

4-4. PIPE PENETRATIONS:
2 inches for tendon and rebar.

5.0 REINFORCING

5-1. All reinforcing bars shall be ASTM A-615 Grade 60, except Grade 40 may be used for stirrups, corner bars and hairpins.

5-2. All tendons shall be 270k grade, 7 wire strand, 1/2 inch diameter, U.N.O., grea sheathed with a continuous extruded plastic sheathing.

5-3. Anchorage system shall be a monostrand unbonded tendon anchorage utiliz wedge plate and a two piece wedge as manufactured by a P.T.I. approved man

5-4. All post-tensioned tendons and anchors shall conform to the requirements of "P.T.I. Guide Specification For Post-Tensioning Materials." Post-tensioned ten to be P.T.I. factory certified.

5-5. PARTIAL STRESS: all tendons to 16.5 kips (or half of final jacking force) 24 t after concrete placement.

5-6. FULL STRESSING of all tendons to 33 kips 7 to 10 days after concrete place

5-7. The first tendon in the slab shall be a maximum of 14 inches and a minimum from the outside form. Tendons not d mensioned on plan to be equally spaced.

5-8. (1) #3 x 24 inches x 24 inches corner bar requi red at all exterior corner's top Deepened beams to have corner bars with diameter equal to horizontal steel at horizontal bar.

5-9. At plumbing stacks, add #3 bars x size of opening plus 16 inches to be place inches beyond per meter of opening (not req'd. If cables are partial stressed - s

6.0 PLAN VARIATIONS

6-1. All depth dimensions of beams are minimum unless intact rock is encountere depth. Inspector may approve beams cont nuously on rock to minimum beam d inches. Deepen beams where required by site conditions at least 6 inches into v or to rock unless deep beam detail applies.

6-2. When PI is 38 and greater and trees are within 15 feet of foundation, consult "Policies Concerning Trees" latest revision.

6-3. Should conditions arise that are not covered by details on this plan, contact E once for additional instructions.

6-4. In areas to receive tile, we recommend installing 6x6x1.4x1.4 WWF 1 1/2" be concrete surface and bedding the tile on a bond breaker to prevent shrinkage c reflecting through the tile.

6-5. HARD POINTS. If the depth of underslab clean fill at any beam intersection (depth, not from beam bottom), exceeds 48 inches, place hard points through the of 12 inch diameter pre-formed or drilled, concrete piers. All beams to have ten steel. (If hardpoint depth exceeds 6'-0" from top of slab reinforce w/ (4) #4 vert. @ 24" O.C.) If total underslab fill exceeds 12 feet, contact Engineer.

7.0 TREE POLICY. APPLIES TO P.I.'S = 38 AND GREATER

7-1. TREE WITHIN 5 FEET FROM FOUNDATION
a. Add 20'-0" of section 3 steel - center on tree in exterior beam only. OR
b. Deepen beam 24" into existing soil for 20'-0" - exterior beam only.

7-2. TREE 5 TO 15 FEET FROM FOUNDATION.
a. Add 20'-0" of section 3 steel - center on tree in exterior beam only, OR
b. Deepen beam 12" into existing soil for 20'-0" - exterior beam only

7-3. Add 6" wide trench 24" into existing grade 20'-0" long centered on tree and f un-reinforced concrete.

OPTIONAL PROVISIONS TO BE ENFORCED, IF CHECKED:

☐ FILL (UNAPPROVED): The fill material on this site is unsuitable to support a slab-on-ground foundation. The fill must be penetrated by all grade beams extend a minimum of 6 inches into virgin soil. As an alternative, see HARD note. Based on the soils investigation, unapproved fill appears to be appr deep.

☐ FILL (APPROVED): The fill material is acceptable to support a slab-on-groun foundation. Construct exterior grade beams 6 inches into approved fill. "A Fill" is fill that has been approved by MLAW, based on proper exploration, inspect on by an agency acceptable to MLAW.

PRE-
POST-
POUR

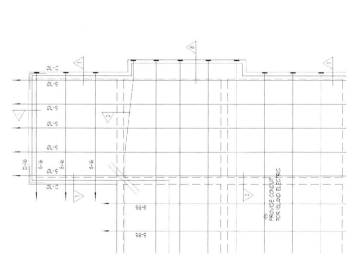

PROVIDE CONDUIT
FOR ISLAND ELECTRIC

FIGURE 8.4

TYPICAL FOUNDATION PLAN NOTES AND SHORT SPECIFICATIONS FOR RESIDENTIAL OR LIGHT COMMERCIAL CONSTRUCTION

8.13.7 Foundations not designed by an engineer.

Foundation plans not designed by an engineer may be used by following the provisions of the local building code. This is appropriate if the buildings are small, such as small-footprint single-family residences, duplexes, or apartments not exceeding two-stories. Small commercial structures may similarly be constructed without engineered foundations provided the code is followed. These provisions are all appropriate if there is a history in the area of no foundation problems with this type of construction and the construction is in general conformance with local practice. These conditions and the need for engineered plans must be determined by the building official at any given locality.

A foundation not designed by an engineer would not be acceptable for major commercial structures, monumental structures, places where large numbers of people congregate, or places where there is a high dollar value of investment and subsequent tax value involved. The prime function of the building codes is to ensure health and safety of the human population, but building codes aside, it is important that buildings of permanent value are constructed to enhance the community and to maintain the local tax base. Certainly all major structures such as buildings, bridges, dams, treatment plants, lift stations, air fields, and major roadways should have engineering consideration including geotechnical studies and engineering design.

8.14 TEST QUESTIONS

MULTIPLE CHOICE

1. What are three things that should be in the soils report (select three):
    a. structural drawings of columns
    b. traffic studies
    c. identification of the client
    d. results of field and laboratory investigations
    e. tide tables
    f. recommendations

2. Generally, if the proposed construction is reasonably heavy and fine grain soils exist at a high water content (select one):
    a. the site may be muddy
    b. foundation settlement is likely
    c. drainage is not needed
    d. the site should be abandoned

3. Shrink-swell activities of a soil are usually more significant (select one):
    a. for heavily loaded structures
    b. in the winter
    c. for lightly loaded structures
    d. when there is a nearby electro-magnetic field

4. A pile driven into a silt/clay soil will nearly always (select one):
    a. cause the soil to become stronger after driving
    b. cause the soil to become weaker after driving
    c. buckle or shatter
    d. cause water to erupt at the ground surface

5. The two basic types of foundations are (select two):
    a. wide
    b. narrow
    c. shallow
    d. deep

6. Selection of the type of foundation should consider the function of the structure along with (select one):
    a. what the contractors like to construct
    b. the sensitivity of the structure to movement
    c. impact on appraised value
    d. proposed landscaping

7. In clay soils, a safe bearing value in most shallow to medium depth footings will be close to (select one):
    a. the unit weight of the soil
    b. water content of the soil
    c. the unconfined compressive strength of the soil
    d. footing depth in feet

8. In a sand deposit supporting a foundation with a width of 10 feet and N-value = 30, the allowable load to cause one inch of settlement is (select one):
    a. one ton per square foot
    b. three tons per square foot
    c. 4.8 tons per square foot
    d. 30 tons per square foot

9. Considerations of construction approaches contained in the geotechnical report could include the need for (select two):
    a. de-watering excavations
    b. job-site traffic control
    c. the need for casing piers
    d. consideration of the qualifications of the job superintendent

10. An attempt to utilize soil reports that were not done on the site, a long way from the proposed construction, is an indication that (select one):
    a. this is a wise example of saving the owner's money
    b. a dangerous situation is evolving
    c. the architect really knows the area soils
    d. there is no reason to be concerned

11. Without a formal geotechnical report (select one):
    a. the owner can save some costs
    b. the structural engineer will know what to do since his standard plans cover all conditions
    c. some knowledge of soil conditions that exist should be known to the foundation designer
    d. the design process goes faster

12. Professional design organizations involved in the preparation of plans normally should (select one):
    a. keep their names, addresses, and telephone numbers off the plans to avoid being drawn into a lawsuit
    b. provide full contact information on the plans to encourage good communication
    c. avoid providing too much technical information
    d. destroy plans as soon as a project is complete

13. State laws typically require that engineer-prepared plans have (select one):
    a. the official seal of the state
    b. the engineer's professional seal and signature and the date of the release of the plans
    c. a secrete bar code known only to the engineer
    d. the name of the architect fully credited with the plan preparation

14. If the foundation designer feels that it is necessary to deviate from the geotechnical engineer's recommendations (select one):
    a. the designer should contact the geotechnical engineer for a supplemental report
    b. the designer should acknowledge that it is generally not a good idea to get a supplemental report from the geotechnical engineer agreeing with the alternate approach since this will produce added work for the geotechnical engineer
    c. the designer can use his own judgment
    d. the designer should ask a likely contractor for his opinion

Chapter 9

EXCAVATION AND GRADING

Site earthwork

Chapter 9

EXCAVATION AND GRADING

9.1 What Are Excavation and Grading?

The first step in constructing something on a site will be to prepare the site in rough fashion to receive the construction. This involves excavation, which is the removal of earth by human operations. Earth excavation is generally performed by mechanical equipment but occasionally by manual labor.

Grading is involved in moving earth around on the site, shaping it to conform to the site plan, and producing the rough grades of the constructed product. Grading is generally separated into rough grading and fine grading. Fine grading is also called finish grading. Rough grades are done by earthmoving equipment such as bulldozers, wheeled scrapers, or motor graders and generally shape the site, level the building pads, and arrange for the major surface drainage features. At this point utilities and any underground drainage pipes are usually installed.

Finish grading is reserved until near the end of the project and is closely associated with landscaping. Paving will need finish grading of the subgrade before the pavement sections are placed. Section R403.1.7 of the 2012 IRC, Sections 1804 and 1808.7 of the 2012 IBC, and Appendix J of the 2012 IBC discuss excavation, grading, and fill.

Section 1804.4 of the 2012 IBC discusses grading and fill in flood-hazard areas in an attempt to ensure that fill will not slump during floods and that the earthwork will not increase the design flood elevation or create other water hazards.

9.2 Definitions Relating to Excavating and Grading

Backfill – Fill placed behind retaining walls, behind basement walls, or within utility trenches. The word backfill implies that a previous excavation was done and that the fill is being placed "back." However, backfill may also be a specified, imported material called "select fill."

Bedrock – Sound rock, also known as "ledge rock." It may be covered by surface weathering profiles or alluvium (Qal) deposits. Bedrock should not be confused with boulders.

Bench – A level area, either excavated or naturally occurring, on a sloping site.

Berm – A low levee constructed of earth that is intended to divert surface water, provide a landscape feature, or provide noise protection.

Bluff – A geologic condition also known as a cliff. A bluff is frequently formed by either large-scale faulting or erosion on a large scale, which leaves a nearly vertical face of rock or earth.

Borrow – Earth transported to a site for the purposes of fill or special construction, originating offsite or somewhere else on the construction site. In many parts of the country rural road sections have a "bar ditch," which is the ditch on either side of the roadway from which the roadway embankment soil was excavated, and which then serves as a drainage ditch. The term "bar ditch" is a colloquialism for "borrow ditch."

Compaction – The increase in density of a soil material usually by mechanical means, although hand tampers are also used in confined spaces. The earth is packed together tightly to expel the air voids, producing a heavier unit weight, higher strength, and stability. Compaction could also occur in nature as a result of overburden compressing lower soils, expelling water over an extended time period. In natural deposits this is more likely to be called consolidation.

Creep – A very slow movement down slope of near-surface soil and loose rocks that will eventually produce damage to constructed facilities located on the surface. Creep is usually seasonally cyclic, moving more in wet periods and slowing or stopping during dry periods. Creep is also seen in snow packs on hillsides and, in both snow and in earth, if creep becomes significantly large it will herald an avalanche or landslide.

Erosion – Involves the loss of soil or rock material due to the actions of water, ice, or wind. Erosion can also be applied to constructed works such as concrete structures if sufficient erosional forces have been at work.

Fill Lift –Uncompacted material that is placed in layers when fills are being constructed. These layers are typically 9 to 12 inches thick prior to compaction and will be reduced in thickness by about one-third after compaction. The use of thin lifts is important in compacting soil to ensure that the compacting forces are able to operate on the full depth of loose soil to provide the required densification. Heavy-duty compaction equipment may be used on lifts deeper than 12 inches.

Grade – Grade is a term commonly used to describe either the initial or final elevation of the ground surface. The grade prior to construction would be called existing grade or original grade, and the finished grade would be the desired final grade of the site in accordance with the plans, both represented by contours or spot elevations on topographical maps. The term "grading" refers to the processes of getting a site configured to the grades required by the plans.

Hydraulic Mulch – Applying a sprayed mixture of water, fertilizer, mulching materials, and seeds to provide a quick grass start and protection of newly-graded raw earth slopes. Some building departments require hydraulic mulch to be in place or even for there to be grass growing on raw earth slopes prior to the final release of the building or site permit.

Mudflow – A river of soil, water, rock, and other debris. It is too soft to walk on and will rapidly flow as a fluid, moving from higher elevations to lower, and it can be a nearly unstoppable force of destruction. Mudflows are triggered by heavy rain, usually in mountainous or hilly areas. The runoff has picked up a considerable erosional load often involving freshly placed earth from construction sites.

Over-Excavation – A condition in which the bearing soil to support a footings is removed deeper than the designed bottom of the footing. The over-excavation must be backfilled with density-controlled soil or lean concrete to achieve the correct elevation for the footing. The procedure to be used is typically described on the structural plans or in the specifications.

Proof Rolling – Rolling the subgrade of a fill, compacted fill lifts, or pavement layers with a heavy, loaded, rubber-tired vehicle to detect soft or unstable areas that require remediation. Usually two complete overlapping passes are used. An inspector will walk behind the vehicle to detect excessive yielding (about 1/10 inch). Proof rolling tests the entire area covering the areas between density tests.

Pumping Soil – Soil that is too wet and undulates during compaction. The soil must be dried for effective compaction or replaced with drier stable material.

Rip Rap – Placing of stone or thin concrete on a sloped earth surface to protect against erosion. It is also a term used to describe protection from breaking waves or running water on shorelines.

Slope – A generic term for a ground surface, either natural or man-made, that is not level. The term is also used to describe the amount of the change in surface elevation divided by horizontal distance. For example, a slope of 1v:20h would be a slope of a 1-foot fall in 20 feet or 5 percent.

Topography – A geographical term that includes the physical attributes of the land, including its slopes, hills, and valleys. A topographic map is one that is frequently needed in site planning and design and contains an illustration of the land surface, typically by contour lines. Contour lines are lines of equal elevation, and the closer together they are the steeper the slope and vice versa.

Well Points – These are used to temporarily lower the water table so excavation can proceed. Well points are similar to water wells with screened casing and pumps to remove water to a desired level.

9.3 Grading and Drainage Plans

Grading and drainage plans are used on nearly every construction project. They are usually prepared by the site civil engineer and show existing land surfaces, usually as dashed contour lines, and the proposed or finished land surfaces, generally with solid contour lines. Sometimes spot elevations are indicated where more precision is needed. The plans show the earth work to be done, including the surface-drainage features and frequently the sub-surface-drainage or storm-sewer system. The grading and drainage plan set also include calculations for the contributing drainage areas and an analysis that shows the amount of storm runoff being contributed at various parts of the site. This portion of the plan is valuable for plan checking of runoff to prevent flooding or to identify possible erosional areas. Plans may show earth swales or valley gutters. Valley gutters are lined with concrete because of the higher water velocities and the danger of erosion. The drainage plan set may also have calculations that illustrate and justify the size of the pipes, inlets, and outlet structures of the storm-sewer system.

Drainage is an important feature of these plans, and the designer must show how the rain water will be conveyed off the site and discharged in a proper and legal manner. In most jurisdictions it is not legal to discharge more flow across property lines than existed before construction or concentrate runoff flow and discharge it onto neighboring property without making provisions for dealing with this water. Sometimes this requires the acquisition of easements to permit the construction of drainage swales or pipe systems. Runoff entering the site from uphill areas offsite must also be considered in the drainage analysis.

Usually the proposed buildings or other structures are shown in these plans, and finished elevations, such as the finished floor, sidewalks, driveways, and other constructed items should be included. The finished elevations adjacent to structures are important because they control possible accumulation of water that could flood into the structure. Section R401.3 of the 2012 IRC and Section 1804.3 of the 2012 IBC contain information about drainage near structures. Figure 9.1 illustrates a typical grading and drainage plan for part of a housing subdivision.

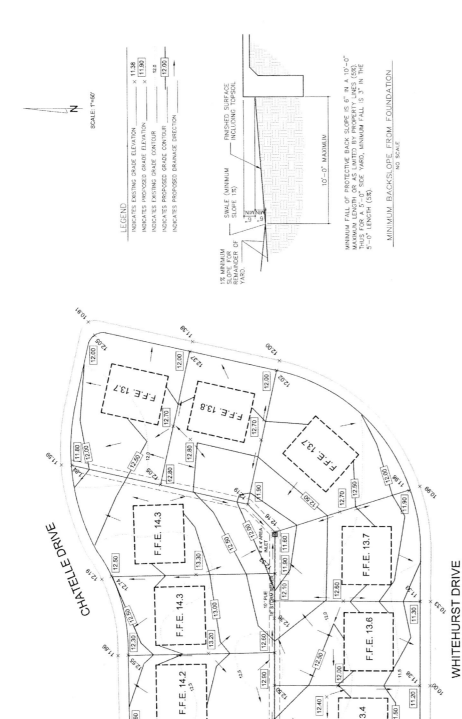

LEGEND

INDICATES EXISTING GRADE ELEVATION	× 11.38
INDICATES PROPOSED GRADE ELEVATION	× [11.90]
INDICATES EXISTING GRADE CONTOUR	12.0
INDICATES PROPOSED GRADE CONTOUR	[12.00]
INDICATES PROPOSED DRAINAGE DIRECTION	

SCALE: 1"=60'

SWALE (MINIMUM SLOPE 1%)

FINISHED SURFACE INCLUDING TOPSOIL

1% MINIMUM SLOPE FOR REMAINDER OF YARD.

10'-0" MAXIMUM

MINIMUM FALL OF PROTECTIVE BACK SLOPE IS 6" IN A 10'-0" MAXIMUM LENGTH OR AS LIMITED BY PROPERTY LINES (5%). THUS FOR A 5'-0" SIDE YARD, MINIMUM FALL IS 3" IN THE 5'-0" LENGTH (5%).

MINIMUM BACKSLOPE FROM FOUNDATION
NO SCALE

SITE GRADING AND DRAINAGE PLAN FOR WOOD HILL, BLOCK B

JOB NO. 210000.500
DATE: AUGUST, 2005

CHATELLE DRIVE

BERWICK DRIVE

WHITEHURST DRIVE

F.F.E. 13.7
F.F.E. 13.8
F.F.E. 13.7
F.F.E. 14.3
F.F.E. 14.3
F.F.E. 14.2
F.F.E. 13.7
F.F.E. 13.6
F.F.E. 13.4

4' x 4' AREA INLET
10' P.U.E.
18" STORM SEWER

FIGURE 9.1
TYPICAL GRADING AND DRAINAGE PLAN

85

9.4 Excavation

Excavation to remove earth can be done using a variety of equipment ranging from a small hand-operated shovel to some very large pieces of equipment. Equipment commonly used for these operations include bulldozers, pan scrapers, wheel or ladder trenchers, backhoes, front-end loaders, draglines, and motor graders. Bulldozers and pan scrapers are used for large-area type excavation, while backhoes and trenchers are used for smaller areas that require more precision, including utility trenches and footings. Draglines are frequently used when the excavation is done under water as in a gravel pit; they are also used in other processes involving wetland construction. Motor graders are used for fine grading and are used on roadway or parking lot construction to finish the grades to very precise tolerances. Section 9.12 illustrates and provides information about various types of excavation and grading equipment.

Section 1804.1 of the 2012 IBC prohibits excavation that would remove lateral support from foundation elements without providing underpinning or other footing protection and Section 1803.5.7 requires a geotechnical investigation of this situation. This seems like a common-sense requirement, but it is often violated with major consequences.

9.5 Fill

9.5.1 Definition of fill. Fill is any soil material that has not been placed by nature. It can generally be divided into three types: engineered fill, forming fill, and uncontrolled fill.

9.5.2 Engineered fill. Engineered fill is that which has been designed by an engineer to act as a structural element of a constructed work and has been placed under engineering inspection, usually with density testing. Engineered fill may be of at least two types. One type is "embankment fill," which is composed of the material found on the site, or imported to no particular specification, other than that it be free of debris and trash. Embankment fill can be used for a number of situations if properly placed and compacted.

"Select fill" is the second type of engineered fill. The term "select" simply means that the material meets some specification as to gradation and PI, and possibly some other material specifications. Examples of select fill are crushed limestone, specified sand or gravel, crusher fines, or other soil that meets the job specification requirements. Select underslab fill is frequently used under shallow foundations for purposes of providing additional support and stiffness to the foundations or for replacing a thickness of expansive or very soft soil. Section 1803.5.8 of the 2012 IBC requires a geotechnical investigation and report if compacted fills over 12 inches in depth are to support shallow foundations.

Engineered fill should meet specifications prepared by a qualified engineer for a specific project, which include requirements for material properties, placement, geometry, compaction, and quality control. Figure 9.2 shows proper compaction procedures on a site. The fill is placed in 8-inch to 12-inch layers, compacted to 6-inch to 8-inch thicknesses, and tested for field density to see if it meets specifications. Figure 9.3 illustrates the use of a nuclear density device to test for proper field density and compaction. Engineered fill should be considered a structural component of the site work and foundations, and treated with the same inspection and quality control as reinforced concrete or any other structural element.

9.5.3 Compaction control. Control of field compaction is accomplished by comparing the density of the soil in the field to a maximum laboratory density.

FIGURE 9.2
MOTORIZED COMPACTOR ROLLING FILL IN A
CONTROLLED SETTING TO PRODUCE
"ENGINEERED FILL"

FIGURE 9.3
FIELD DENSITY TEST USING NUCLEAR
DENSITY METER

The maximum laboratory density is obtained by pounding a specimen of soil in a mold 4 to 6 inches in diameter in layers according to test specifications as shown in Figure 9.4. Four to six molds of the same soil are mixed with different amounts of water and subjected to the same laboratory compaction procedure. The molding water content and the dry density of each specimen is recorded and plotted to form a moisture density plot, also known as a Proctor curve.

The soil will compact more readily and become denser as water content is increased until a certain point is reached, at which point the density will decrease as the soil approaches saturation. A Proctor curve or moisture-density relationship curve is a plot of dry density versus molding water content, as seen in Figure 9.5.

From this curve the maximum laboratory dry density of the material in pounds per cubic foot can be determined. This does not mean this is the maximum density that could be obtained, but this is the maximum that can be obtained under the particular compaction specification that has been selected. By comparing the field and laboratory dry densities, the testing laboratory can note a percentage of field compaction compared to the maximum laboratory density and decide if this percentage is too low to meet job specifications.

Typically earthwork specifications for new work call for 90, 95, or 100 percent of maximum laboratory density based on one of the compaction standards discussed below.

The two most commonly used compaction standards are the Standard Proctor and the Modified Proctor. ASTM D – 698 describes the Standard Proctor laboratory compaction procedures. The Modified Proctor compaction procedure uses about four times the laboratory pounding energy to develop the moisture-density curves and results in somewhat higher maximum laboratory densities and lower optimum water content. This procedure was originally developed for compacting fills and bases for airfield pavements and is now typically also used for industrial and major roadway pavements. It is more difficult for a contractor to achieve these densities in the field; however, this compaction standard is often used for building sites. This procedure is described in ASTM D–1557. The 2012 IBC J 107.5 calls for fills to be compacted to 90 percent of Modified Proctor (ASTM D-1557).

The molding water content from the Proctor curve that is close to the maximum laboratory dry density is known as the optimum moisture content and will indicate the field water content at which field compaction can be accomplished with the least amount of rolling. A range of field water contents a few percentage points dry to a few percentage points wet of optimum is often specified for new work to help insure adequate compaction.

Some soils compact easier than others. Material that will compact readily is a typical roadway flexible base of crushed stone containing coarse sizes graded down to silt sizes. High clay content soils are more difficult to compact, and silty soils may be very difficult to compact. If during compaction the water content of the soil is too low, compaction cannot be achieved, and it is desirable to have the moisture content of the soil within a few percentage points of the optimum moisture content for that particular material. If the soil is too wet, the clays and silts will react badly and will not compact, but simply roll around and pump

under the wheels of the compaction equipment. At this point additional rolling will not provide additional compaction because the soil is past the optimum water content and it is in a saturated condition. For these conditions, the soil will have to be scarified and dried out or replaced with a more suitable soil.

FIGURE 9.4
TECHNICIAN PERFORMING THE PROCTOR TEST TO OBTAIN MAXIMUM LABORATORY DENSITY OF A SOIL MATERIAL

When a clay soil is subjected to fine grading, for example to create a subgrade of a pavement, the moisture content should be nearly optimum to avoid tearing and cracking of the clay surface under the action of the grader blade.

Compaction does not change the specific gravity of the soil grains, but simply reduces the void spaces between them. Compaction forces air out from between the grains, causing the whole mass to become denser. Since a soil mass consists of soil grains, water, and air, when nearly all the air is forced out by compaction the soil mass becomes saturated and becomes unstable. This is illustrated by the falling density seen on the right side of a Proctor curve as the molding water is increased as illustrated in Figure 9.5.

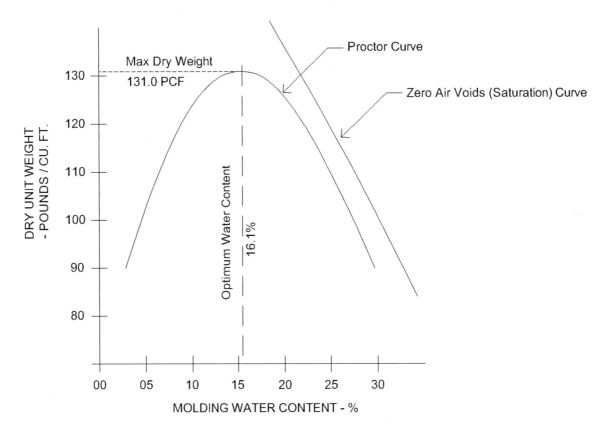

FIGURE 9.5
EXAMPLE OF A PROCTOR CURVE

If settlement of man-made fills is suspected, the procedure is to obtain the density of the fill in place by nuclear testing methods or other field density procedures and compare them to the possible maximum laboratory densities. This assumes that the fill is clean of trash, organics, and nested boulders.

Example of testing fill for possible settlement:

If the field dry density is 87.1 pcf and the maximum laboratory dry density from the Proctor curve is 105.6 pcf, the percent compaction in the field is 87.1/105.6 = 82.5 percent. This percentage would be considered too low, and the fill will likely settle unacceptably.

9.5.4 Forming fill. Forming fill is that which is often used under residential or light commercial shallow foundation slabs to act as a part of the concrete forming. It is normally not expected to be heavily compacted, and a designer should not rely on this material for support. The only requirements of using forming fill are that this material be non-expansive, clean, work easily, and stand when cut. If forming fill happened to be properly compacted and inspected in accordance with an engineering specification, it could be engineered fill.

9.5.5 Uncontrolled fill. Uncontrolled fill (non-engineered fill) is fill that has been determined to be unsuitable (or has not been proven suitable) to support a foundation or other engineering works. Any fill that has not been approved for its purpose by a qualified geotechnical engineer in writing should be considered uncontrolled fill. Uncontrolled fill may contain undesirable materials, and/or it has not been placed under compaction control. Some problems resulting from uncontrolled fill include gradual settlement, sudden collapse, attraction of wood ants and termites, corrosion of metallic plumbing pipes, and in some rare cases, site contamination with toxic or hazardous wastes.

9.5.6 Fill placement. Fill is placed by equipment similar to that described above for excavation depending on the situation. Both excavation and fill operations may be serviced by dump trucks to move the material further than can be conveniently moved by dozers or scrapers.

Fill should be compacted, especially when it is used as a structural element to support a foundation or roadway (Section 1804.5 of the 2012 IBC). Compaction is expedited by having the moisture content of the soil brought close to the optimum water content by the use of water trucks and mixing. See 9.5.3 for more details about compaction. When water is added to a loose fill, a disk plow may be used to mix the moisture throughout the lift to be compacted. If the soil is too wet to properly compact, it may be dried out by aeration using a disk plow or other equipment to turn the soil and loosen it to permit evaporation to occur.

Prior to fill being placed, the existing natural ground must be prepared by having 6 to 12 inches of topsoil, grass, trees, and brush removed, including roots. The exposed soil is then loosened by plowing and re-compacting with density testing. Sloping fill surfaces may require benching to prevent a steeply sloping surface between the fill and the natural ground, which could form a slip plane. The need for this process depends on the soil materials involved and the height of the fill, but typically when the slope exceeds 1v:10h, benching is probably a good idea. (1v:10h is an example of a standard method of describing a slope. The "v" is the vertical component and the "h" is the horizontal component.) The 2012 IBC J 107.3 requires benching for fills on existing grades of greater than 1v:5h.

Granular materials generally compact easier than fine-grain soils. Well-graded granular material, such as well-graded gravel or crushed stone, will compact quite readily with minimum effort. Clay soils may require more compactive effort to break down the hard clods and exclude the air. If uncontrolled fill or backfill of clay soil is placed, the clods will arch against each other and trap large air voids. With time, the clay clods will soften due to buildup of moisture and gradually collapse, possibly producing settlement of up to several feet.

Lift thickness to be employed during compaction is typically specified in the plans or specifications book. Generally, clay soils should have loose lift thicknesses of 8 or 9 inches, which will compact to about 6 inches.

Gravelly soils or crushed stone road base may be successfully compacted with thicker base lifts of 10 to 12 inches, compacting to about 8 inches. The thickness of lifts is also a function of the weight of the compaction equipment. Vibratory rollers are widely used for clay or granular soils, but are very effective for granular soils.

Boulders in fill should be considered carefully, especially if footings or stiffener beams or drilled piers or pilings are to be placed through the fill later. Appendix J, Section J107.4 of the 2012 IBC prohibits hard material greater than 12 inches in any dimension being placed in fills.

Sheepsfoot rollers have been used for a long time, especially for clay soils. The original type of "sheepsfoot" had an enlarged base on 6- to 8-inch steel spikes set around a steel drum. Because of difficulty in keeping clay from being trapped above the enlarged bases, recent equipment uses a tapered enlarged spike with a "chevron" shaped foot for clay. Padfoot rollers have round or square pads set around a steel drum and are most effective for granular material. Properly used, the sheepsfoot or tapered foot should penetrate nearly the entire thickness of the loose lift and compact the lift from the bottom up. If the sheepsfoot or tapered foot does not "walk out" of the lift being compacted after two or three passes, the soil is probably too wet for proper compaction.

9.5.7 Foundation considerations for fill.
The presence of fill on a site can significantly impact the performance of a foundation placed on it. If the fill is engineered fill, the impacts may be positive. If, however, it is uncontrolled fill, significant settlements and major failure of the foundation could result. Foundations should not be supported by uncontrolled fill.

Fill may be placed on a site at various times, and if it has been placed prior to the geotechnical investigation, the geotechnical engineer should note fill in the report. Fill may exist between borings or be undetected during the geotechnical investigation for a variety of reasons. The investigation will be more accurate if the borings are more closely spaced. Occasionally, fill is placed after the geotechnical investigation is completed, and it may not be detected until foundation construction is started.

If unacceptable fill is discovered later in the construction process, for instance by an inspector after a foundation is completely set up and awaiting concrete, great expense may be incurred by having to remove the reinforcing and forms to provide penetration through the fill. Therefore, it is important that such materials be identified and a strategy developed for dealing with them early on in the construction process.

9.5.7.1 Settlement.
Uncontrolled or non-engineered fill can produce settlements under the imposed weight of the structure or its own weight; the settlements can range from a few inches to as much as 10 feet, depending on the nature and depth of the fill. Often uncontrolled fill is not all soil materials, but may include organics, trash, junk, and even toxic waste. If the site was used as an official or unofficial landfill (dump), anything could be found there, most of which will probably consolidate with time and may produce poisonous gas and dangerous leachate discharge. If a fill is found by a site investigation to be non-engineered or cannot be tested after the fact to show that it is equivalent to an engineered fill, the fill should not be relied upon to support a foundation. The choices would be to remove the fill and replace it in a proper manner (including removal of trash, junk, or organics and using controlled compaction) or to extend deep foundation elements to a more stable material below. Neither of these approaches is popular with builders, owners, or contractors since each entails additional and often unexpected costs and schedule increases.

Landfills or dumps that might produce poisonous gases or toxic leachate should not be covered with habitable structures, regardless of the foundation type. Occasionally systems are installed using sealing and venting to permit construction of buildings over dumps. The author has seen several such schemes fail after a few years requiring abandonment of the buildings and resulting in major legal action.

To establish foundations on non-engineered fill, penetration through the fill is required to take the load-supporting elements of the foundation below the unreliable zone. Penetration could be accomplished by deepened stiffener beams, spread footings, or piers depending on the depth and the economics of the situation. Generally, drilled piers or piles are most cost effective once the fill to be penetrated exceeds about 4 feet, but this depends on the foundation engineer's judgment and local practice. Floor systems must be structurally designed to span between the support points, not relying on the unacceptable fill. The non-engineered fill is assumed to offer no long-term support but it can be used to form the surface concrete.

There are several proprietary procedures available to deal with deep uncompacted fill such as compaction grouting, dynamic compaction, and deep vibratory compaction. The use of these techniques is beyond the scope of this book and will always require the utilization of a specialist to determine if they are appropriate, determine the methodology for using them, and to confirm their effectiveness.

Fill on a construction site should always be considered unreliable until it is proven satisfactory. Proof could consist of testing reports of density as the material was placed from bottom to top, all under the review of a geotechnical engineer. The reports must include the engineer's statement that the material is suitable for supporting the proposed foundations or pavement. Alternatively, the fill can be tested after the fact by the use of test pits, borings, proof rolling, or other procedures, also under the supervision of a geotechnical engineer, followed by the engineer's statement. If neither of these conditions can be met, the fill must be assumed to be incapable of supporting the foundation, and other alternatives should be pursued. Organic materials in landfills or "dumps" can take up to 10 years or more to decompose and settle, and sometimes surprises come along years after construction.

Figure 9.6 illustrates the process of "proof rolling," which assists in acceptance of uncontrolled fill. Proof rolling is only useful for depths of 2 to 3 feet. Deeper fills will need to be partially removed to facilitate valid proof rolling. Soft spots that are located are excavated and reconstructed as required until the area passes the proof-rolling test. This process is also very often done for pavement or foundation pad construction as a final "proof" of the compacted subgrade and base layers. Two overlapping passes of heavy wheel loads are required.

FIGURE 9.6
PROOF ROLLING TO ASSIST IN ACCEPTANCE OF UNCONTROLLED FILL AND VERIFICATION OF OVERALL STABILITY IN NEW WORK. A FULLY LOADED DUMP TRUCK OR WATER TRUCK MAY BE USED. THE INSPECTOR WALKS NEAR THE LOAD VEHICLE TO NOTE SOFT OR YIELDING AREAS. OVERLAPPING WHEEL PATHS ARE USED.

9.5.7.2 Bearing. Sometimes engineered fill is used to increase the bearing value of spot or strip footings. The footings rest on the top of the fill, which may be placed in strips or pads, and the bottom of the fill is constructed with a wider contact area, reducing the bearing pressure on the soil below. Generally, the contact pressures between the footing bottom and the top of a compacted granular fill can be higher than in naturally occurring soft soil. These procedures should be designed by a qualified engineer, based on valid field information.

9.5.7.3 Heave. Heave due to expansive clay can be reduced under a shallow foundation by removing the upper clay soils to a depth of 2 feet to 10 feet and replacing them with an inert fill material (typically silty sand, gravel, or crushed stone), properly compacted. If the "remove and replace" option is selected for an expansive soil site, care should be taken to ensure that random water from rain or other sources is not trapped within the excavated clay area and held in the more porous granular material. This is known as the "bathtub" effect and can cause water to pond on expansive clay soils for extended periods of time, thereby creating more heave than would have occurred without the "remove and replace" procedure. One method for dealing with this is to put a drainage blanket and subsurface drains under the fill pad, with the drains exiting by gravity. It is also helpful to either pave over with sidewalks or pavement, or place 12 inches of compacted clay fill over the top of the replacement fill where it is exposed outside of the perimeter of the foundation. This is to reduce the amount of water that could penetrate into the replacement fill. However, it should be assumed that water will eventually penetrate from some source and will need to be removed by the drain system described above. Figure 9.7 illustrates the use of a "remove and replace" procedure with the subdrains described.

9.5.7.4 Constructability. The nature of fills on the site may affect the constructability of the project. If the fill material is relatively fine grained with very low plasticity, such as silt or "screenings" from a crusher, it is possible that stiffener elements or other vertical parts of the foundation cannot be formed against the fill since the material may cave in, especially after rains. Such materials are very difficult to compact and may become unstable under compaction rolling outside of a very narrow range of water content. If silts are exposed to rain after compaction, they can become saturated easily and will become unstable under construction traffic. Clean sands almost always require compaction by vibration.

If deep foundations are to be employed, such as piers or piles, the presence of boulder fill or trash fill may be important to note since such fill could materially affect the ability to install these types of foundations. Large rock boulders in a fill of any depth can cause the price of a drilled pier shaft to rise astronomically because of the difficulty of drilling shafts through this material. It is also obvious that piles cannot penetrate such material readily because of the danger of a pile splintering or deflecting on a large boulder. If a drilled pier, reinforced concrete pile, or steel pile is to penetrate a landfill that contains organic material, consideration should be given to the possibility of acid attack on concrete or steel. The acid can be generated from decomposing organic material. The concrete may have to be designed using more cement or a Type-V Cement. In addition, the cover of concrete over reinforcing steel may need to be increased. Steel piles may not be suitable in such fills.

If a site for building foundations or pavement is unstable, special measures must be employed. Such conditions could include organic soil or a high water table. These conditions should normally be addressed in the geotechnical report, site plans, or foundation plans. If the situation has not been addressed in the project documents, the geotechnical engineer should be asked to recommend solutions.

For organic soil, the best procedure is to remove the soil down to stable soil and replace it with select fill such as gravel or crushed stone, placed with density control. Alternatively, the building foundations may be extended through the unstable soil using piers or piles. Such changes to the foundation scheme should be coordinated with the structural engineer.

Sites that are unstable due to a high water table can possibly have the water table lowered, or open-graded coarse rock can be placed into and on the soil to bridge the area and provide a stable platform to construct the foundations. The water table can be permanently lowered using French drains. Ditches can be used to remove the water as a temporary measure during construction. Extreme situations may require the use of well points. It is always best to have the geotechnical engineer provide recommendations as to procedures and testing to verify stability.

9.6 Utility-Trench Backfill

When utility trenches are placed beneath foundations or pavements, special compaction procedures should be followed.

Utility and storm-drain trench backfill under foundations or pavements should be placed in lifts, with the moisture content adjusted to near the optimum and compacted by trench rollers or tampers. The practice of "water ponding" or "water compaction" is a discredited procedure that will generally result in future settlement of the trench backfill. Trench backfill under pavements is frequently brought up by the compaction process described above, with the upper foot or two beneath the pavement receiving special treatment in the form of concrete or compacted crushed limestone. Settling of utility-trench backfills under newly constructed pavements is one of the prime causes of pavement damage. Even worse is a utility trench cut through an existing pavement, in which case special precautions including a concrete cap underneath the pavement structure are frequently used.

Backfilling trenches with controlled compaction of earth is a difficult and slow process. Flowable fill or CLSM (Controlled Low-Strength Material) is a useful fill material to completely backfill a utility or storm-drain trench and consists of a mixture of sand, cement, fly ash, air entrainment chemicals, and water placed by pumping or direct dumping from a ready-mix truck. This has the advantage of speeding construction and providing a guaranteed resistance to settlement once the cementitious mixture hardens. Usually flowable fill is batched to produce a fairly low strength, on the order of 100 to 500 psi, instead of 3000 psi for typical concrete. This is because it may be necessary to open the trench again in the future, and it will be better for the utility not to have to air hammer out 6 feet of concrete to reach the pipe. One disadvantage of flowable fill is that it costs more than simply compacting earth. However, the speed of backfill is often an advantage, especially when crews are working in a street or parking lot that is in constant use. See Section 1804.2 and 1804.6 of the 2012 IBC.

In open country in which the utility or storm drain is not to be covered by pavement or a foundation, nominal compaction can typically be employed. This is less expensive than the controlled compaction process; however, it is a good idea for 1 to 2 feet of earth to be mounded above the finished surrounding grades to provide for future settlment. Specifications typically call for the contractor to return in

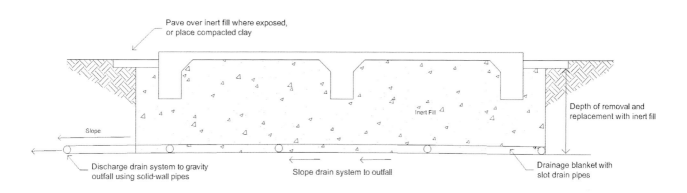

FIGURE 9.7
REMOVAL AND REPLACEMENT TECHNIQUE WITH SUBDRAIN

three months to adjust any settlement that might or might not have taken place and thus present a uniform appearance.

9.7 Retaining-Wall Backfill

Retaining walls come in a variety of styles, and all of them are designed to retain earth on an abrupt change of elevation. Compacted backfill behind retaining walls is good practice, but over-compaction can cause excessive lateral force to be exerted against the retaining structure beyond the normal soil forces for which the wall is designed. Backfilling behind retaining walls should be done carefully so the back-drainage features that may be constructed with the wall system to prevent the buildup of excess hydrostatic pressure behind the wall are not disrupted. Basement walls should be considered retaining structures, and similar rules for backfilling should apply. See Section 1804.2 of the 2012 IBC.

Basement walls often require bracing until the structure is built over the basement to keep them from caving inward under the lateral force of the backfill or the compaction activities; the structure itself would then take up the job of keeping the walls apart.

9.8 Valley Fill

When it is necessary for a valley to be filed to provide a level area for development or construction, several precautions should be observed. The first precaution is that all topsoil, organics, trees, and brush should be removed, and the remaining subgrade soil should be scarified and compacted. Benching may be necessary depending on the slopes. Most valleys have been formed by the action of running water, and it is reasonable to assume that water may run again in its previous pathway between the subgrade and the new fill. Percolating water in these fills can soften the soil and cause erosion where the water outcrops at the toe of the fill, leading to massive settlement and collapse. Therefore, it is usually a good plan for a subdrain system to be placed on the surface of the former valley floor leading to an outfall pipe that has an armored and controlled exit point at the base of the fill.

High fills in valleys sometimes require the construction of special earth embankments at the downstream edge of the valley to act as an earth dam to contain the fill that is to follow. The construction of these embankments should be properly engineered and constructed with density control. It may be necessary for a different type of soil to be used in order to obtain sufficient strength.

9.9 Compaction Control to Reduce Swelling

Compaction control to reduce swelling is a technique used to control swelling soils by moisture conditioning. In this procedure the soil is excavated to a specified depth and recompacted maintaining the fill moisture content from 1 percent to 5 percent above the Proctor optimum moisture content. The increased moisture will reduce the soil's "thirst" for more water and thus reduce future swell. A lower range of density is required, for example 92 percent to 97 percent of the maximum laboratory reference density instead of 100 percent. If this technique is used, the allowable bearing values of the soil will be reduced, and this should be considered in the foundation design.

9.10 Hydroconsolidation and Hydrocollapse

Hydroconsolidation may occur in fills composed of hard clay particles, even if subjected to compaction, because the hard clay particles will hold up until they become moistened over time. The particles will then slip past each other and form a more compact mass, expelling air that was originally present. This can often be seen in the settlement of utility-trench backfill and more significantly in the settlement of deep valley fills or high embankments. Structures built on these fills can be seriously damaged by the settlement. One solution is to compact the soils on the wet side of the Proctor optimum moisture.

Hydrocollapse is a more dramatic form of settlement, and this term is used for a naturally-occurring soil that has an open structure held up by cementation between particles. This condition is often found in desert or semi-arid environments. The soil deposit is strong as long as it is kept dry. If the soil becomes saturated due to surface construction, drainage, plumbing leaks, or a septic field, the bonds can soften, and sudden collapse is possible. The movement may be several feet. The best solution is for the engineer to identify the depth and extent of such areas, and the earth to be excavated and recompacted before construction or the excavation replaced with select fill.

9.11 Appendix J of the 2012 IBC

Appendix J of the 2012 IBC is devoted to code grading requirements. Jurisdictions that adopt the IBC should also adopt Appendix J if it is to be made part of the local ordinance. This appendix lists permitting procedures (and exemptions) that include requirements for a site plan showing grading and a soils report. Section J104.4 requires a liquifaction study for sites with maximum mapped earthquake spectoral response accelerations of 0.5g or more for short periods per Section 1613.

Section J105 calls for general inspections per Section 110 of the code and the special inspections found in Section 1705.6 for work performed under a grading permit, if required by the building official.

Other items covered in this appendix include maximum slope on cut or fill surfaces, placement of fill and compaction (J107); setbacks from property lines (J108); slope protection (J108.3); drainage, terracing, and surface drainage provisions (J109); and erosion control (J110).

9.12 Earthwork Equipment

Not shown in the illustrations of earthwork equipment is a walk-behind plate compactor. This is a steel plate equipped with a motorized vibrator that can be handled by personnel walking behind it and guiding its movements. The vibration permits the plate to be readily moved to compact soil in a small area not accessible to larger compaction equipment. Because of the low contact pressures of the plate compactors, the process should be applied to thin soil lifts. A variation of the walk-behind plate compactor is a larger version that can be attached to a small tractor.

A recent development becoming more widely used is "Intelligent Compaction," which uses force sensors and GPS mounted on vibratory rollers to map a full record of the compaction achieved over the entire fill site. This is used to demonstrate "quality control" by the contractor and provide data over 100 percent of the area, while fill density testing is limited to specific locations. A testing laboratory will follow up with standard field densities providing "quality assurance."

FIGURE 9.10
BULLDOZER. THESE ARE USED TO PERFORM SURFACE EXCAVATION OR LOCAL DEEP EXCAVATIONS AS WELL AS SPREADING SOIL OR SELECT FILL IN LIFTS. BULLDOZERS CAN ALSO BE USED TO ASSIST OTHER EQUIPMENT SUCH AS WHEELED SCRAPERS BY PUSHING OR PULLING.

FIGURE 9.8
LARGE BACKHOE. THESE ARE USED FOR UTILITY TRENCHES OR LOCALIZED EXCAVATION AND LOADING HAUL TRUCKS.

Photo: Caterpillar

FIGURE 9.11
WHEEL TRACTOR-SCRAPER. THIS IS A CATERPILLAR 627H, A TWIN-ENGINE OPEN-BOWL SCRAPER WITH A CAPACITY OF 24 CU. YDS. AND A LOADED TOP SPEED OF 33 MPH. IT CAN CUT (EXCAVATE) A 10-FOOT-WIDE BY 12-INCH-THICK LAYER OF SOIL, MOVE IT TO ANOTHER AREA, AND DISCHARGE THE SOIL IN A CONTROLLED LAYER OF FILL. OTHER METHODS OF EXCAVATING, FILLING, AND MOVING EARTH INCLUDE BULLDOZERS, EXCAVATORS, AND HAUL TRUCKS AND MOTOR GRADERS. HOWEVER, THE WHEEL TRACTOR-SCRAPER IS MORE EFFICIENT ON LARGE JOBS.

FIGURE 9.9
SMALL BACKHOE. THESE ARE USED IN SITUATIONS SIMILAR TO THE LARGE BACKHOE BUT FOR SMALLER WORK.

FIGURE 9.12
WATER TRUCK USED TO ADJUST SOIL MOISTURE TO APPROXIMATELY OPTIMUM FOR EFFECTIVE COMPACTION OR CONTROL DUST ON A JOB SITE.

Photo: Caterpillar

FIGURE 9.13
VIBRATORY PAD-FOOT ROLLER. THIS IS A CATERPILLAR CP64 USED TO COMPACT CLAY OR GRANULAR SOILS. IT IS MOST EFFECTIVE FOR GRANULAR OR SANDY SOILS. PADS CAN BE ROUND OR SQUARE. THE TAPER OF THE PADS REDUCES SOIL FLUFFING DURING PAD UP-LIFT. THE DRUM WIDTH IS 84 INCHES AND RATE OF PRODUCTION OF COMPACTED ROAD BASE OR SOIL IS ABOUT 500 CU. YDS. / HR. VIBRATION FREQUENCY AND ROLLER WEIGHT CAN BE ADJUSTED TO PROVIDE OPTIMUM RESULTS. VIBRATION FREQUENCY IS STANDARD AT 1914 VPM, UP TO 1800 VPM FOR LARGE MODELS. VIBRATION AMPLITUDE CAN BE EITHER 0.070 INCHES OR 0.035 INCHES PRODUCING CENTRIFUGAL FORCE OF 63,300 LBS. OR 31, 600 LBS. FOR HEAVY CLAY, THE TAPERED-FOOT, 4-WHEEL COMPACTOR OR THE OLDER SHEEPSFOOT ROLLERS ARE MOST EFFECTIVE.

Photo: Caterpillar

FIGURE 9.14
VIBRATORY DRUM ROLLER. THIS CATERPILLAR CS64 IS SIMILAR TO THE PAD-FOOT ROLLER, BUT USES A SMOOTH ROLLER DRUM. IT IS OFTEN USED TO SMOOTH AND FINISH A SUBGRADE, THE SURFACE OF PAVEMENT BASE COURSES, OR TOP OFF BUILDING FOUNDATION PADS.

Photo: Caterpillar

FIGURE 9.15
FOUR-WHEEL SOIL COMPACTOR. THIS IS A CATERPILLAR 815F SERIES 2 SOIL COMPACTOR. IT IS PRIMARILY USED FOR COMPACTION AND IS VERY USEFUL FOR HEAVY CLAY. THE TAMPING WHEEL TIP CONFIGURATION HAS A "CHEVRON" SHAPE TO INCREASE MAXIMUM GROUND PRESSURE. SIMILAR TO THE TAPERED-PAD MACHINES, COMPACTION IS FROM THE BOTTOM OF THE LIFT, AND THE TIPS WALK OUT WHEN COMPACTION IS ACHIEVED. THE BLADE ALSO PERMITS SPREADING FILL. DEPENDING ON SLOPE, BLADE LOAD, AND GEAR SELECTION, THE COMPACTOR CAN OPERATE AT 4 TO 12 MPH. OPERATING WEIGHT IS ABOUT 46,000 LBS.

Photo: Caterpillar

FIGURE 9.16
PNEUMATIC ROLLER. THE TOTAL WEIGHT OF THIS MACHINE RANGES FROM 19,000 LBS. EMPTY TO 55,000 LBS. WITH MAXIMUM BALLAST LOAD. BALLAST CONSISTS OF WATER, STEEL BOLT-ON WEIGHTS, STEEL WEIGHTS PLUS WATER, AND STEEL PLUS WET SAND. THE PNEUMATIC TIRES PRODUCE AN ECCENTRIC OR WOBBLE MOTION THAT SERVES TO KNEAD THE SOIL. TIRE AIR PRESSURE CAN RANGE FROM 35 TO 110 PSI. THE RUBBER TIRES CAN HAVE THE CONTACT PRESSURE ADJUSTED BY AIR PRESSURE AND CHANGE OF BALLAST. GROUND CONTACT PRESSURES CAN BE VARIED FROM 71 TO 83 PSI AVERAGE. MAXIMUM CONTACT PRESSURES CAN RANGE FROM 130 TO 140 PSI SINCE THE CONTACT PRESSURE IS NOT UNIFORM. ROLLING SPEED IS ABOUT 5 MPH, AND COMPACTION WIDTH IS 90 INCHES. THE BEST USE OF THIS MACHINE IS TO COMPACT SANDY CLAY AND CRUSHED STONE PAVEMENT BASE. HEAVY CLAY IS BEST COMPACTED USING A TAPERED-FOOT COMPACTOR. SMALLER PNEUMATIC ROLLERS ARE USED FOR HMAC COMPACTION.

Photo: Caterpillar

FIGURE 9.17
MOTOR GRADER. THIS CATERPILLAR 14M MOTOR GRADER WEIGHS 47,000 LBS. AND HAS A 14-FOOT BLADE, WHICH CAN BE ANGLED UP TO 65°. IT IS EQUIPPED WITH A RIPPER ON THE REAR THAT CAN RIP TO A DEPTH UP TO 16 INCHES. DURING HEAVY GRADING THE MACHINE MOVES AT UP TO 6 MPH. FOR FINISH GRADING, LOWER GEARS ARE USED FOR MAXIMUM PRECISION, AND SPEEDS ARE ABOUT 3 MPH. THIS MACHINE IS USED FOR MOVING EARTH FOR PAVEMENT SUBGRADE PREPARATION, GENERAL AREA GRADING, AND FINISH GRADING OF BASES. THE RIPPER IS USED TO BREAK UP HARD EARTH OR ASPHALTIC PAVEMENT PRIOR TO BLADING. MOTOR GRADERS ARE USED FOR BUILDING-SITE PREPARATION, ROAD CONSTRUCTION, DITCH CONSTRUCTION, AND SNOW REMOVAL. BLADING ON SIDE BANKS UP TO 2H:1V CAN BE DONE.

Photo: Caterpillar

FIGURE 9.18
RECLAIMER/STABILIZER. THIS MACHINE MAY BE USED TO CUT OUT OLD ASPHALTIC PAVEMENT AND BASE MATERIAL, MIX IT WITH LIME OR CEMENT, AND LEAVE A LIFT OF STABILIZED MATERIAL FOR COMPACTION. IT IS NORMALLY USED IN PAVEMENT WORK TO PRODUCE SUB-BASE OR BASE LAYERS.

Photo: Wishek Manufacturing, LLC

FIGURE 9.19
DISC PLOW. THIS IS AGRICULTURE-BASED EQUIPMENT USED IN CONSTRUCTION FOR SCARIFYING SOIL FOR MOISTURE CONDITIONING, BLADING, AND COMPACTION. DIFFERENT MODELS ARE AVAILABLE WITH 10-FOOT OR 16-FOOT CUTTING WIDTHS AND WEIGHTS OF 15,000 TO 21,000 LBS. THE PLOW IS PULLED BY A TRACTOR OR BULLDOZER. THE HORSEPOWER REQUIRED VARIES WITH THE TYPE OF SOIL AND WORKING DEPTH.

9.13 Grading Control

Earthwork on almost every type of construction project requires excavating and filling of soils in order to shape the site to the desired conditions. On all but the very smallest projects, the finished elevations of the site are dictated by the site grading and drainage plan prepared by the site civil engineer. To achieve the desired elevations in the field, the contractor must follow cutting and filling instructions indicated on what are called "cut stakes" placed by a surveyor. They can also be called "cut and fill stakes." Typically, this process involves the placing of a ground-level wooden hub or a splay of plastic fibers set with a nail at the current ground level and a $1/4$-inch by $1^1/_2$-inch wooden lath about 3 or 4 feet above the ground level driven into the ground next to the hub point. The surveyor determines at each hub point whether the soil at that location requires excavating or cutting, or filling. The amount of cut, for example, would be indicated on the lath using terminology such as C 2.5 feet or in the case of fill it could say F 1.2 feet. The earthwork equipment operator would note this and proceed to cut or excavate to the requested amount or fill to the requested amount. Intermediate measurements may be made with a survey instrument to see how close to accomplishing the cut fill requirements have been made. In some cases the cut and fill stakes may be offset, and the notation reflects the offset distance such as "10 feet left" or "10 feet north." Often the ground-level hub and the lath are preserved on a large earthwork site, with the excavating or filling being done around it and keeping the reference elevation intact for as long as possible before it is lost to the grading process. At the conclusion of the rough grading involving mass movement of the soils, it may be necessary to reset the cut-fill stakes and hubs for fine grading purposes.

For finish grades, sometimes "blue tops" are used. These are wooden 2-inch by 2-inch hubs or flagged nails set exactly at the desired finish elevation below a nearly completed surface. The motor grader operator blades off the soil or pavement material layer to match the "blue tops."

More modern techniques, which are often used, provide that earthmoving equipment, especially for fine grading, be equipped with a receiver that is impacted by a level rotating laser beam. The receiver determines at what elevation the laser beam intersects it and can indicate to the operator how much more cutting or filling is necessary to achieve the desired grades at any particular point. Recent developments use GPS (Global Positioning System) to control cutting and filling on major earthwork projects. This method locates the equipment coordinates on the site and advises how much cutting or filling is necessary at that location. The site plans must be installed in an office computer system and downloaded to the equipment modules.

Cut stakes are also used in utility construction, such as sanitary or storm sewers or large-diameter potable water lines that are to be constructed to a defined slope and elevation. Often utility-trench stakes are set on each side of the trench out of the way of excavating equipment and are used to set a string line or straight edge at a level position between the opposite sides of the trench. The string line is then used as a reference at a known elevation for a tape or level rod to determine when the bottom of the trench is at the desired elevation at that point. The same procedure can be used to develop drainage channels or ditches, but this is more frequently done with cut-fill stakes as described previously. In all cases of earthwork control an allowance must be made for a dressing of top soil or pavement sections that will be placed over the earthwork.

On smaller jobs, earthwork contractors may do the grading, including cutting or filling, by "eyeball" and then check their work with a surveyor's level when they think they are close to being finished with the job.

9.14 TEST QUESTIONS

MULTIPLE CHOICE

1. In most jurisdictions it is (select one):
   a. legal to concentrate runoff flow and discharge it onto a neighboring property without making provisions for dealing with this water
   b. not legal to do the above
   c. legal to concentrate runoff flow and discharge it onto a neighboring property if you advise the property owner of that

2. Water ponding or water compaction is (select one):
   a. a procedure which will generally result in very little settlement of trench backfill
   b. a discredited procedure that will result in future settlement
   c. a procedure that will assist in establishing vegetation

3. Hydrocollapse is a condition (select one):
   a. that is caused by a failing dam
   b. that may be found in desert or semi-arid environments
   c. that is the result of a ruptured water main

4. Grading is generally separated into (select two):
   a. rough grading
   b. fine grading
   c. hillside grading
   d. motor grader work

5. The sections in the 2012 IRC and the 2012 IBC that discuss excavation, grading, and fill are (select three):
   a. 1805 and 1808.7 of the IBC
   b. R403.1.7 of the IRC
   c. Appendix G of IBC
   d. Appendix J of IBC
   e. 1804 and 1808.7 of the IBC

6. Section 1804.4 of the 2012 IBC discusses grading and filling in flood hazard areas. Things to be aware of for these conditions (select two):
   a. fill will not slump during floods
   b. fill will not increase the flood elevation
   c. the earthwork should provide an escape route
   d. the fill will not trap animals

7. In earthwork, a level area, either excavated or naturally occurring on a sloping site, is called a (select one):
   a. plain
   b. berm
   c. bench

8. A nearly vertical face of rock or earth is called a (select one):
   a. rockfall
   b. bluff
   c. rill
   d. berm

9. A definition of compaction is (select one):
   a. the increase in density of a soil material usually by mechanical means
   b. reduction of a specific gravity of soil grains
   c. hydrodynamic consolidation

10. The term grading refers to processes of (select one):
    a. smoothing a surface
    b. getting a site configured as required by the plans
    c. ripping hard rock

11. A slope of 1v:10h would describe a one foot fall in (select one):
    a. 100 feet
    b. 1000 feet
    c. 10 feet

12. Contour lines are lines of (select one):
    a. even curves
    b. shape factors
    c. equal elevation

13. Compaction is expedited by bringing the moisture content of the soil close to the (select one):
    a. saturated water content
    b. optimum water content
    c. shrinkage limit
    d. relative humidity

14. What does the abbreviation CLSM stand for? (select one):
    a. Cold Layer Surface Material
    b. Climatic Latitude Site Maps
    c. Controlled Low-Strength Material

15. What is the definition of uncontrolled fill? (select one):
    a. fill that is placed wildly
    b. fill that has not been proven suitable and reported in writing
    c. fill placed at night
    d. fill containing cobbles

16. What are the problems that could result from uncontrolled fill in a construction project? (select three):
    a. poor appearance
    b. settlement
    c. poisonous gas
    d. cost overruns

17. Uncontrolled or non-engineered fill is assumed to offer (select one):
    a. partial long-term support
    b. no reliable support
    c. as much support as any other fill

18. What is a definition of fill? (select one):
    a. any soil that has not been placed by nature
    b. any soil that contains both sand and clay
    c. any soil that does not contain vegetation

Chapter 10
SOIL AND SEISMICS

10.1 Why Classify Site Soils for Seismic Conditions?

10.2 Site Classification Procedure (2012 IBC Section 16.3.2)

10.3 Test Questions

The photo was taken by the NSF-Sponsored Geotechnical Extreme Events Reconnaissance (GEER) Association. Courtesy of Jonathan D. Bray, University of California, Berkeley.

Earthquake damage

Chapter 10

SOIL AND SEISMICS

10.1 Why Classify Site Soils for Seismic Conditions?

In a seismically-active area, different soil materials react differently to the same amount of earth shaking. For example, a rock site may have the same earthquake energy put into it as a less stable site, but because of its rigidity it will undergo much smaller displacements and therefore shake a structure built on it much less. On the other hand, a soft, deep deposit of soil will react in the extreme to the same amount of energy and cause major lateral movements that will be transmitted to a structure built on it. This is the same principle as striking a hammer on the side of a bowl filled with hardened concrete and striking the same hammer stroke on a bowl filled with Jell-o. The Jell-o will quiver and shake whereas the concrete will not. These are extreme examples, and soils will range in between these examples. Some types of soils may actually liquefy under earthquake shaking and lose nearly all their strength (Sections 1803.3, 1803.5.11 and 1803.5.12 of the 2012 IBC).

Liquefaction can permit large areas of soil to flow, taking structures with them and causing ground settlement of many feet. Soil zone liquefaction due to earthquake forces often causes slope failures, but massive slope failures can occur even without liquefaction just through the horizontal and vertical force increases. Site soils should be classified, and their reaction should be determined under anticipated earthquake forces. Mitigation measures could include seismic structural design, avoidance of soils that can liquefy, and strengthening of slopes to resist earthquake forces.

10.2 Site Classification Procedure (2012 IBC Section 16.3.2)

Site class A, B, C, D, E and F definitions are to be made using Chapter 20 of ASCE 7.

Table 10.1 is taken from Table 1613.5.2 in the 2006 IBC and indicates the requirements for classifying soil conditions for use in determining site properties for various seismic active areas. The table in the 2006 IBC is the same as Table 20.3.1 in ASCE 7.

TABLE 10.1
(TABLE 1613.5.2 IN 2006 IBC)
SITE CLASS DEFINITIONS

SITE CLASS	SOIL PROFILE NAME	AVERAGE PROPERTIES IN TOP 100 feet, SEE SECTION 1613.5.5		
		Soil shear wave velocity, $\bar{v}_s$, (ft/s)	Standard penetration resistance, $\bar{N}$	Soil undrained shear strength, $\bar{s}_u$, (psf)
A	Hard rock	$\bar{v}_s > 5,000$	N/A	N/A
B	Rock	$2,500 < \bar{v}_s \leq 5,000$	N/A	N/A
C	Very dense soil and soft rock	$1,200 < \bar{v}_s \leq 2,500$	$\bar{N} > 50$	$\bar{s}_u \geq 2,000$
D	Stiff soil profile	$600 \leq \bar{v}_s \leq 1,200$	$15 \leq \bar{N} \leq 50$	$1,000 \leq \bar{s}_u \leq 2,000$
E	Soft soil profile	$\bar{v}_s < 600$	$\bar{N} < 15$	$\bar{s}_u < 1,000$
E	—	Any profile with more than 10 feet of soil having the following characteristics: 1. Plasticity index $PI > 20$, 2. Moisture content $w \geq 40\%$, and 3. Undrained shear strength $\bar{s}_u < 500$ psf		
F	—	Any profile containing soils having one or more of the following characteristics: 1. Soils vulnerable to potential failure or collapse under seismic loading such as liquefiable soils, quick and highly sensitive clays, collapsible weakly cemented soils. 2. Peats and/or highly organic clays ($H > 10$ feet of peat and/or highly organic clay where H = thickness of soil) 3. Very high plasticity clays ($H > 25$ feet with plasticity index $PI > 75$) 4. Very thick soft/medium stiff clays ($H > 120$ feet)		

For SI: 1 foot = 304.8 mm, 1 square foot = 0.0929 m², 1 pound per square foot = 0.0479 kPa. N/A = Not applicable

Using the table:

1. The values of V_s (shear velocity), N (N value from the Standard Penetration Test), and S_u (undrained shear strength) for site classes A, B, C, D, and E are weighted averages for the upper 100 feet of the soil profile. The bar over the table headings indicates an average. The averaging procedures are discussed in Section 20.4 of Chapter 20 of ASCE 7. In these procedures Σ means to sum up or add. The "n" and "i=1" indicates the range of the layer count. The depths of each different layer (d_i) are added up (totaling 100 feet) on the top of the equation, and the depths of each layer within the 100 feet divided by the value of that layer are added up on the bottom. After division of the top by the bottom, the weighted average is obtained for use in the table. Any one of the average properties—$\overline{V}_s$, $\overline{N}$, or $\overline{S}_u$—may be used to assign the site class. The symbol ≤ means equal to or less than, and ≥ means equal to or greater than the numerical value shown.

2. Site class A may be assigned based on a report by a qualified professional that shear-wave velocities are 5000 ft/sec or greater. These values may be based on site-specific tests or by relating the site to similar rock profiles with known values. Higher shear-wave velocities indicate a more rigid material, which is less prone to near-surface movements under earthquake shaking.

3. Site class B may be assigned based on the procedures for Site Class A for shear-wave velocities of 2500 to 5000 ft/sec for sites with moderate fracturing and weathering. Softer and more highly fractured rock shall have a site-specific test for shear-wave velocities or be assigned to Site Class C.

4. Site classes A and B cannot be used if there is more than 10 feet of soil below the bottom of the footings above the rock.

5. Site class C, D or E may be assigned using any of the three average values of $\overline{V}_s$, $\overline{N}$, or $\overline{S}_u$ as shown in the table. If the soil is a soft clay with thickness greater than 10 feet with $\overline{S}_u$ less than 500 psf, water content greater than 40 percent, and PI greater than 20, the site will be Class E.

6. Site class F should be assigned for any of the following conditions: liquefiable soils, quick and highly sensitive clays, collapsible soils, greater than 10 feet of peat and / or highly organic clay, greater than 25 feet of clay with PI greater than 75, or 120 feet or more of soft / medium stiff clay with $\overline{S}_u$ less than 1000 psf.

7. Seismic Design Categories C through F require a site-specific geotechnical investigation. (2012 IBC 1803.5.11 and 1803.5.12)

After the structural engineer is provided the site class by the geotechnical engineer, he or she must apply other provisions of Section 1613 of the 2012 IBC and chapters 12 and 22 of ASCE 7 to determine structural loadings from earthquakes and perform the design. Discussion of these procedures is beyond the scope of this book.

10.3   TEST QUESTIONS

MULTIPLE CHOICE

1. A rock site and a site with a soft, deep deposit of soil are subjected to the same amount of seismic energy (select one):
   a. they will react the same
   b. will shake less if a rock site
   c. the soil site will shake less

2. In Table 1613.5.2, site classes C, D, and E use the weighted averages for the values of V_s, N_s and S_u for the upper (select one):
   a. 100 feet of the soil profile
   b. 50 feet of the soil profile
   c. 10 feet of the soil profile

3. In Table 1613.5.2, the symbol ≤ is used to mean (select one):
   a. left reference
   b. equal to or less than
   c. equal to or greater than

4. Soils that lose nearly all their strength under earthquake shaking are said to undergo (select one):
   a. shear loss
   b. liquefaction
   c. increase in water content
   d. excessive vibrations

5. Site Class A may be assigned based on a report by a qualified professional that soil shear-wave velocities are (select one):
   a. 500 ft/sec or greater
   b. 2500 ft/sec or greater
   c. 5000 ft/sec or greater
   d. 10,000 ft/sec or greater

6. If the soil is a soft clay with thickness greater than 10 feet, S_u less than 500 PSF, moisture content greater than 40 percent, and PI greater than 20 the site will be Site Class (select one):
   a. A
   b. B
   c. C
   d. D
   e. E
   f. F

7. If a project needs to be established in a risk area of Site Class F (select one):
   a. a formal geotechnical report and analysis of the soil is not required
   b. a formal geotechnical report and analysis of the soil is needed
   c. all that is needed is a study of geologic maps
   d. the structural engineer and architect can provide designs without a site study

Chapter 11
SPREAD OR STRIP FOOTINGS

11.1 Bearing Capacity Check

11.2 Settlement Check

11.3 Footing Design

11.4 Installation

11.5 Test Questions

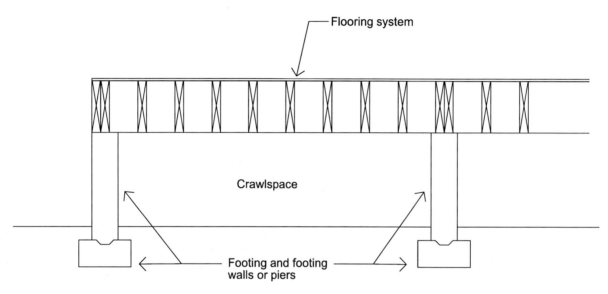

A footing foundation

Chapter 11

SPREAD OR STRIP FOOTINGS

(SECTION R403 OF THE 2012 IRC AND SECTIONS 1808 AND 1809 OF THE 2012 IBC)

11.1 Bearing Capacity Check

The design of spread or strip footings, which are a common type of shallow foundations, require a check of the loads applied to the footing divided by the allowable bearing capacity in pounds per square foot. This will yield the area of the footing required to avoid local shear failure immediately beneath the footing.

Example: A load applied to a square spread footing of 20,000 pounds using a safe bearing capacity of 2,500 pounds per square foot will yield a required contact area of the footing of 20,000 divided by 2,500 = 8 square feet. If the footing is square, it will be slightly more than 2.8 feet or about 2 feet, 10 inches in each direction.

A strip footing is usually employed to handle a line load such as a wall, and the applied load per running foot would also be divided by the safe bearing capacity to provide the width of the footing. The footing should not be too narrow, and a minimum width of 18 inches to 24 inches is suggested to provide stability. Footings supporting light frame construction may be as narrow as 12 inches per Table R403.1 of the 2012 IRC and Table 1809.7 of the 2012 IBC.

11.2 Settlement Check

If the footings are large and closely spaced for heavy structures, their stress fields may overlap at a short distance below the ground surface and in effect become a uniform pressure after a certain depth. This pressure should be compared to the allowable pressures for the soil to prevent long-term settlement of the building. A large individual footing should also be checked for settlement since soft materials may be only a few feet below the ground surface. The design should be verified with the recommendations of the geotechnical engineer. Settlement is generally not a problem for strong soils and certainly not for good rock. Settlement of shallow footings on fill requires similar considerations, but it is dependent on the type of material and the compaction that has been achieved and verified.

11.3 Footing Design

Once the sizing of the footing is determined as discussed above, the footing itself will need to be designed as a structural unit. For small footings this may not be important since even plain (un-reinforced) concrete can serve as a small footing. However, for larger footings, 2 feet on a side or larger for example, the concrete must be reinforced with steel to properly resist the bending stresses introduced by the load coming down on the center of the footing. The thickness of the concrete as well as the reinforcement are considered in this design. In addition, punching shear should be evaluated. Normally, spread footings should not

be less than 6 inches in thickness and larger ones can be much thicker depending on load conditions. Concrete is not the only material that can be used for footings. Sections 1809.9, 1809.11, and 1809.12 of the 2012 IBC provide guidance for masonry, steel grillage, and timber foundations. Section 1809.13 of the 2012 IBC requires footings be interconnected by adequate ties in seismic categories D, E, or F.

Footing walls may be analyzed for bearing as a strip footing. See 17.4.1 in this book for a discussion of concrete and masonry walls, including footing walls, relative to lateral forces. Figure 11.1 indicates the use of spread footings for columns and strip footings for foundation walls on a site with shallow rock.

FIGURE 11.1
USE OF SPREAD AND STRIP FOOTINGS
ON A ROCK SITE

11.4 Installation

After the footing is designed and the contractor is at work, the installation of the footing should conform to the intent of the plans to provide proper bearing and construction configuration. The surface upon which the footing rests should be level or sloped not more than 10 percent (1v:10h), or it should be stepped to follow sloping ground. 2012 IRC R403.1.5 and 2012 IBC 1809.3 permit 1v:5h. The surface should be cleaned of loose material, and the material supporting the footing should be intact original material, unless a specified compacted select fill is designed and tested. The length, width, and thickness of the foundation should be in accordance with the plans, and the steel reinforcing should be placed in accordance with the plans. Often steel reinforced footings have L-shaped bars in the center that dowel up beyond the top of the footing to receive the plinth or short column that carries the loads from the structure to the footing. These should be located in the proper position prior to placing concrete. Of course concrete should not be placed in a footing excavation if the excavation has water or mud, and the usual precautions should be taken with regard to proper concrete placement.

Section R403.1.7 of the 2012 IRC and Section 1808.7 of the 2012 IBC discuss the placement of footings relative to clearances and setbacks from slopes. Figure 11.2 is taken from Figure 1808.7.1 of the 2012 IBC (it is also Figure R403.1.7.1 of the 2012 IRC).

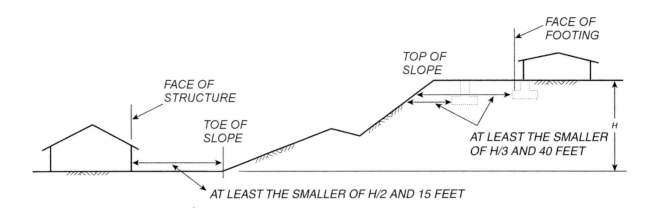

FIGURE 11.2
FOUNDATION CLEARANCES FROM SLOPES
(TAKEN FROM 2012 IBC FIGURE 1808.7.1)

11.5 TEST QUESTIONS

MULTIPLE CHOICE

1. The surface of the soil supporting the footing should be (select one):
   a. intact original soil unless a specified compacted fill is present
   b. compacted fill not over 4 feet thick
   c. any soil surface that is free of boulders or organics

2. Spread or strip footings are common types of (select one):
   a. deep foundations
   b. mat foundations
   c. shallow foundations
   d. hybrid foundations

3. If a load of 25,000 lbs is applied to a square spread footing, utilizing a safe bearing capacity of 2500 psf, the required contact area of the footing will be (select one):
   a. 25 square feet
   b. 100 square feet
   c. 10 square feet

1. Footings and foundations are discussed in (select three):
   a. 2012 IRC Section R406
   b. 2012 IBC Section 1808
   c. 2012 IBC Section 1809
   d. 2012 IBC Section 1807
   e. 2012 IRC Section R403

2. A strip footing is usually employed to handle a line load such as a wall. The applied load per running foot would be divided by a safe bearing capacity to provide the (select one):
   a. length of the footing
   b. thickness of the footing
   c. width of the footing

3. Footings constructed on sloping ground should be stepped if the slope is greater than (select one):
   a. 1v:2h
   b. 1v:20h
   c. 1v:10h
   d. 1v:100h

4. Settlement in footings is generally not a problem for (select two):
   a. strong soils
   b. silty sand
   c. rock
   d. deep sand
   e. gravel with clay

5. In larger footings, concrete must be reinforced with steel to properly resist the following (select one):
   a. wind forces
   b. bending forces
   c. gravity loads

6. 2012 IBC Section 1809.13 requires footings to be interconnected by adequate ties for Seismic Design Category (select three):
   a. A
   b. B
   c. C
   d. D
   e. E
   f. F
   g. G

7. 2012 IBC Figure 1808.7.1 illustrates the placement of footings relative to clearances from (select one):
   a. fences
   b. roadways
   c. power poles
   d. slopes

Chapter 12
PIER FOUNDATIONS

Courtesy of Champion Equipment Sales, L.L.C. and the Trevi Group

A pier drilling rig setting a casing at the Leaning Tower of Pisa

Chapter 12

PIER FOUNDATIONS

(SECTION 1810 OF THE 2012 IBC)

12.1   When to Use Pier Foundations

Piers are useful if the near-surface material does not have sufficient strength or is too compressible to support shallow foundations. Piers can be drilled through this softer material to a depth to carry the building loads to the stronger material. Piers may also be useful in the case of highly expansive soils to establish the foundation into a stable zone below the moving soils near the surface. Sometimes piers are used in place of piles if protection of nearby structures from the effects of pile-driving vibration is important. Generally, much larger loads can be transferred by drilled piers then by piling, provided strong material is available within reasonable distances from the ground surface. Figure 12.1 is a pier drilling rig at work.

FIGURE 12.1
PIER DRILL RIG

12.2   Establishment Depths

Depths of establishment are typically determined by the geotechnical investigation report and may be modified by the foundation designer to adjust for combinations of skin friction and end bearing. Figure 12.2 illustrates the considerations for establishment depth in non-expansive soil and expansive clay.

12.3   End Bearing and Skin Friction

Piers can transmit loads to the soil through end bearing and through skin friction. The amount of allowable end-bearing pressure and the amount of skin friction that can be utilized in design is determined by the geotechnical engineer based on soil types and strength. Figure 12.3 illustrates the considerations of skin friction and end bearing for a straight shaft and a belled pier.

The method of calculation of end bearing and skin friction is shown below:

End Bearing

$P \quad = \pi r^2 Q_a$

$P \quad$ = end bearing force in pounds

$r \quad$ = straight shaft or bell radius in feet

$Q_a \quad$ = allowable bearing in psf

$\pi \quad$ = 3.1416

Example: A pier is 2.5 feet in diameter or has a 1.25 foot radius with Q_a = 4000 psf. Available end-bearing total capacity is 3.1416 × 1.25 × 1.25 × 4000 = 19,635 pounds or 20 kips. A factor of safety of three is already included in Q_a.

Skin Friction

$F \quad$ = π d f divided by factor of safety, where π d is the circumference of the pier

$F \quad$ = available skin friction in pounds per foot of skin friction depth

$d \quad$ = pier shaft diameter in feet

$f \quad$ = ultimate skin friction in pounds per square foot of pier surface. "f" is computed for clay soil using the ultimate soil shear strength times an "alpha factor" of 0.55.

Total skin friction (lbs) = F × depth/factor of safety

Example: A pier of 2.5 feet diameter picks up skin friction in clay along a length of 20 feet. The soil ultimate shear strength (f) is 2000 psf. Available total skin friction is 3.1416 × 2.5 × 20 × 2000 × 0.55 = 172,788 pounds or 173 kips. Applying a factor of safety of three yields a safe total skin friction value of 58 kips.

113

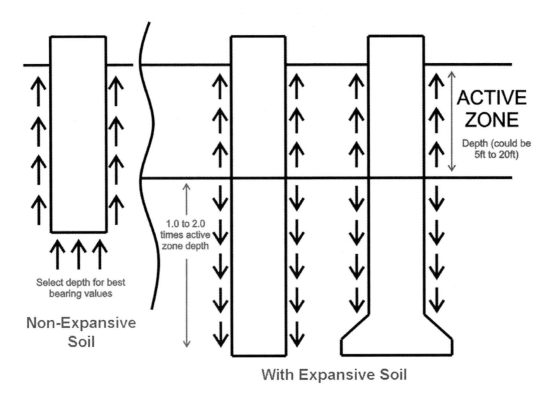

FIGURE 12.2
CONSIDERATION FOR PIER DEPTH ESTABLISHMENT

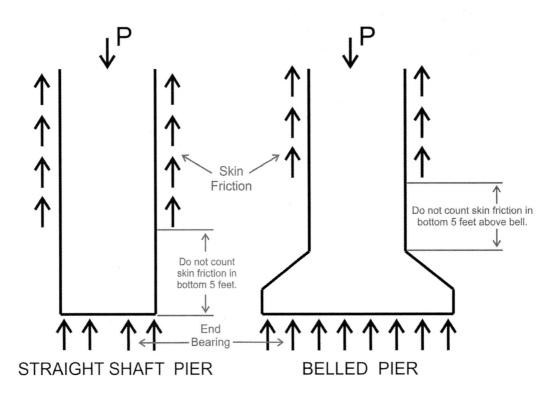

FIGURE 12.3
ELEMENTS IN DESIGN OF PIERS CONSIDERING END BEARING AND SKIN FRICTION

For larger piers, the total of end-bearing force and skin friction should not be just added together to give the total pier capacity. This is because the full end bearing is not realized until 1 to 4 inches of settlement has occurred, depending on diameter, while skin friction is mobilized much sooner. The designer should consider this correction to reduce end bearing for a satisfactory design.

Note: "kips" is often used instead of pounds. A kip is 1000 pounds. For example: 100,000 pounds = 100 kips

Piers may be made deeper to pick up more skin friction and the diameter reduced, or the pier made shallower and the diameter increased to pick up more end bearing. This is a designer consideration and is based on economics. An exception is the use of piers in expansive clays in which a certain depth must be achieved to properly anchor the lower part of the pier in a stable zone to avoid the pier being picked up due to near-surface swelling clay.

12.3.1 Bells or straight shafts. Bells are used to anchor piers against uplift, but only adequate depth and skin friction will do this without getting 1 to 4 inches of movement. Bells were originally used to increase end bearing in weak soils. This is still a good idea. However, the larger the bell, the more movement is required to mobilize full end bearing. This is true for downward loads as well as uplift. Bells are impractical in caving soil unless the pier is cased through to more stable material. In expansive soil, straight shafts established deep enough below the active zone are more stable than shallower belled piers.

12.3.2 Piers and expansive soil. If piers are used to combat the effects of expansive clay soils, the piers must be designed to extend out of the ground and the foundation elements rested on the piers, clear of grade by at least 8 to 12 inches or with a crawl space beneath the floor. This will permit the clay to swell and shrink underneath the foundation, but not contact any horizontal member to cause distress or damage. The expanding clay simply slides up and down on the piers. Figure 12.4 illustrates an inappropriate use of piers with expansive clay.

- For slab-on-ground, the engineer should not connect pier top to slab and expect it not to break if the slab heaves.

- If the problem is settlement, not heave, dowelled connections are acceptable.

12.4 Reinforcement

Once the pier depth and diameter are established, reinforcing steel is supplied for the purposes listed below:

- Full-length steel in piers prevents separation of pier from uplift.

- Steel can increase the load-carrying capacity for heavily-loaded piers established in hard materials.

- Steel can help resistance to lateral bending caused, for example, by downhill creep or lateral structural and earthquake loads.

- Typically 0.75 percent of pier cross-section area is minimum steel. (Example: eight #6 bars in a 24-inch diameter pier). More steel may be required based on uplift calculations.

- Three inches of concrete cover required outside of pier steel where pier is in contact with earth.

Example: to design against piers pulling apart due to swelling clay:

TUF (Total Uplift Force) (lbs) = $\pi \times d \times f \times$ depth of active zone.

Note: $\pi \times d$ is the circumference of the pier. When multiplied by the active zone depth, the total contact area of the pier surface in the active zone is obtained. Multiplying the total contact area by the uplift skin friction yields the total uplift force.

NUF (Net Uplift Force) (lbs) = Total uplift minus dead load on pier.

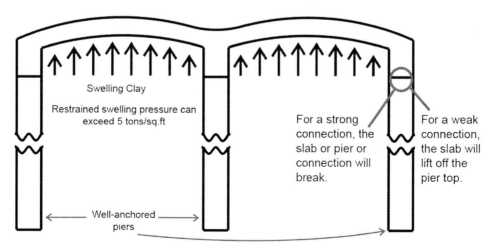

FIGURE 12.4
INAPPROPRIATE USE OF PIERS AND SOIL-SUPPORTED SLAB IN SWELLING CLAY

Steel area needed = net uplift force / 24,000

Where:

π = 3.1416

working stress of steel = 24,000 psi (factor of safety is included)

f = effective uplift skin friction in psf

d = diameter of pier in feet.

Steel area needed divided by selected bar area = number of vertical bars

For a 24 inch (2-foot) pier, 10-foot active zone, f = 2500 psf, TUF = 3.1416 × 2 × 10 × 2500 = 157,080 lbs. For dead load = 30,000 lbs., NUF = 127,080 lbs. Steel area required 127,080 / 24,000 = 5.30 square inches (Use seven #8 bars) (a #8 bar has 0.79 square inches of cross-section area)

In all examples, the soil allowable bearing and skin friction values would be obtained from the geotechnical report.

Figure 12.5 shows a typical pier heavily reinforced for an expansive clay site.

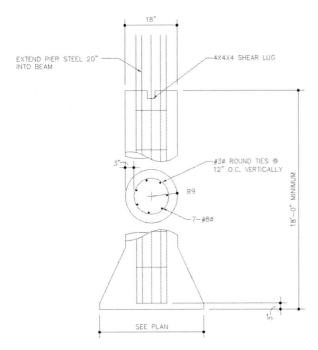

FIGURE 12.5
TYPICAL HEAVILY REINFORCED UNDERREAMED PIER USED IN EXPANSIVE CLAY

12.5   Installation

12.5.1 In rock. Piers installed to reach a rock formation below the ground surface are typically drilled to the surface of the rock using an auger of the proper diameter. The plans typically specify a minimum penetration into the rock, and this penetration can be achieved using either a rock auger, which has tungsten carbide teeth, or a core barrel, which is a cylinder with tungsten carbide teeth on the perimeter. If significant penetration into the rock is required to pick up additional load-carrying capacity from skin friction, attention should be given to the specifications or the geotechnical report recommendations concerning the preparation of the side walls in the rock socket. Sometimes water is utilized in this type of hole, or sometimes seepage water appears. This can mingle with the rock cuttings (typically dust sizes) to form a softened zone that will inhibit achieving the required skin-friction values. The surface of the rock walls may need to be cleaned or scraped to provide fresh rock for contact with the concrete. This is not always the case, and the specifications should be consulted. For heavily loaded piers, which penetrate deeply into the rock for adequate skin friction to add load carrying support, the specifications should be clear, or the designer should be consulted about the condition of the rock wall of the shaft. Careful inspection is required to insure the specified condition is achieved.

12.5.2 In stiff clay. Piers can be readily and economically installed in stiff to hard clay using a clay auger bit of the proper diameter. These types of soil materials would typically stand readily without caving, and rapid progress can be made. These materials also will be able to be underreamed readily. In all cases piers should have a clean bottom without cutting build-up or loose soil chunks, which would adversely affect the bearing capacity of the bottom of the pier.

12.5.3 Boulder deposits. Boulder deposits, either in man-made fill or naturally occurring, are difficult for pier drilling. Sometimes core barrels must be used to cut through the boulders without displacing them into the hole. If boulders are displaced into the drilled shaft, large cavities will result in the pier wall sides, increasing the concrete take and possibly interfering with the skin friction calculations with regard to swelling clay uplift. These types of deposits are difficult to penetrate with piers or piles, and the going may be slow. The inspector must not decide that the bottom of the pier has reached rock when actually it has simply reached a large boulder that may have many feet of softer soil below it. In stiff clay, a drilling rig can put 15 to 20 piers down in a day provided they are of moderate size. However, in a boulder deposit, it would not be unusual to find production down to one or fewer per day.

12.5.4 Water and caving holes. A difficult problem for drilled pier installation is unstable ground including groundwater and holes that cave from the sides. In such situations the holes must be cased using a steel cylinder, sometimes placed through a processed slurry mixture in the pier hole. The slurry, which can be composed of clay or chemical combinations with water, is a viscous fluid that, because of its weight, will restrain the hole walls from caving and keep the cuttings in suspension. The casing can be placed through this material and sealed below the caving area at which point the slurry and cuttings can be removed by the use of dipping buckets or other means. It is important that the casing be properly sealed at the bottom so that the hydraulic head of groundwater does not break through and penetrate into the pier. These foundations can be quite expensive to install. However, this situation is encountered frequently, especially near water when building, for example, bridge piers or structures built close to a lake or river. The steel casing, also known as "temporary steel casing," is normally withdrawn as soon as the concrete is placed and maintained at a level higher than the groundwater head. The casing can then be reused on other holes. Control of the concrete mix is necessary for this procedure to work, and usually a richer cement mix is used with a higher slump to permit the casing to be extracted. If the concrete begins to set up before the casing is extracted, or the concrete is a harsh mix, it may come out of the hole with the casing, bringing the reinforcing cage with it. In many cases this means that the pier will have to be cleaned out and redone. Some specifications permit a slight upward movement of the steel cage when withdrawing the casing, provided that it settles back down into place when the casing is released. The permitted movement is usually on the order of one inch. In these cases the steel casing must be left in place, much to the consternation of the drilling contractor.

Under proper specialist supervision, sometimes the steel casing can be eliminated and the hole simply processed using the slurry. Concrete is then placed by means of a tremie, starting at the bottom of the shaft, forcing the slurry out the top. This procedure is fraught with danger and should only be done by expert contractors with careful inspection.

If a cased hole has severely caving conditions and is quite deep so there is a considerable head of groundwater, care must be taken when withdrawing the casing so the concrete head is sufficient to keep the water and caving soil out of the shaft. Some notable pier failures have occurred because water and mud have intruded into the wet concrete during the withdrawal process, and the intrusion was not noticed. One way to observe possible intrusion is to carefully monitor the volume of concrete take of the shaft and then to monitor the elevation of the concrete surface as the casing is withdrawn. These are difficult inspection procedures. In any event, concrete should not be placed in a pier that has had serious caving, meaning loose soil and water has accumulated in the bottom of the pier shaft. This would totally invalidate the bearing capacity that was assumed in the design.

12.6 TEST QUESTIONS

MULTIPLE CHOICE

1. Drilled piers may be readily used (select two):
   a. to support large loads
   b. if the near-surface material does not have sufficient strength or is too compressible to support shallow foundations
   c. in a boulder deposit
   d. if good bearing soil is more than 250 feet deep

2. In a pier, full end bearing is not realized until _____ inches of settlement has occurred while skin friction is mobilized almost immediately (select one):
   a. 10 to 12
   b. 0.001
   c. 1 to 4
   d. zero

3. For a slab foundation with piers on swelling clay, the pier tops should (select one):
   a. be strongly connected to the slab to resist the swelling clay
   b. extend above ground 8 to 12 inches
   c. be below the ground surface
   d. be at various elevations to avoid concentrating the swell effects

4. Installing drilled piers in stiff clay or in a boulder deposit will (select one):
   a. have the same production rate
   b. have greatly different production rates
   c. be too expensive

5. When withdrawing a temporary steel casing from a deep pier hole, it is important that (select one):
   a. the concrete head is kept at least 10 feet below the groundwater level at all times
   b. concrete be placed after the casing is removed
   c. the concrete head is kept above the groundwater level at all times

6. In non-expansive soils, pier depths are typically determined by the foundation designer (select one):
   a. to allow for reinforcing steel strength
   b. to adjust for combination of skin friction and end bearing
   c. to be at least ten times the pier diameter
   d. to match the drilling equipment

7. If the pier is 3.0 feet in diameter and the soil has a $Q_a = 4,000$ psf, what is the available end-bearing force capacity? (select one):
   a. 12,000 lbs.
   b. about 28,000 lbs.
   c. about 37,700 lbs.

8. 50,000 lbs. = (select one):
   a. 500 kips
   b. 5000 kips
   c. 50 kips

9. In swelling clays, full-length reinforcing steel to prevent separation of the pier due to uplift is typically a minimum of (select one):
   a. 5 percent of the pier cross-section
   b. 10 percent of the pier cross-section
   c. 0.25 percent of the pier cross-section
   d. 0.75 percent of the pier cross-section

10. In designing piers in expansive clay, the abbreviation
    NUF is defined as (select one):
    a. Non-Utilized Force
    b. Net Upward Friction
    c. Net Uplift Force
    d. Normal Underground Force

11. For installing drilled piers into rock, the penetration
    can be achieved using a rock auger or a core barrel,
    both of which have (select one):
    a. diamond cutters
    b. hardened brass teeth
    c. tungsten carbide teeth
    d. self-sharpening steel teeth

Chapter 13
PILE FOUNDATIONS

A village built on piles in a Swiss lake

Archeologists have determined that driven piles have been used since Neolithic times. The lake-dwellers of what is now Switzerland, northern Italy, Austria, and eastern France constructed their villages on driven-pile-supported platforms as a defensive measure. The builders of the ancient village discovered at Wangen, Switzerland, utilized 50,000 piles in the construction of their village. Primitive civilizations around the world have used driven piles and poles to support their homes in water and in flood-prone areas. Predecessors to the Aztecs in the Mexico City area built their sites on lakes in the area on pile-supported mats topped with soil called "Chinampas" or on areas contained by driven piles and filled with soil and stone. Ancient tribes in the Java Sea area built stilted villages. Ancient Chinese engineers built bridges on driven piles.

Prepared by Dr. D. Bossard. From Charles Lyell, Antiquity of Man. Courtesy of Dr. David C. Bossard, 19thCenturyScience.org

Chapter 13

PILE FOUNDATIONS

(SECTION 1810 OF THE 2012 IBC)

13.1 When to Use Pile Foundations

Thousands of years ago a prehistoric people living on a lake in what is now Switzerland pounded poles into the mud at the bottom of the lake to produce a platform on which to build their houses. This is the first known use of piling in construction, although it probably occurred many times previously. As with most things in construction, procedures generally started out by someone trying something, and eventually the process became codified. Equations were written by engineers to attempt to explain what was going on and to extend the procedure for designing other similar construction. Pilings are most useful when the soil is unstable and soft and where excavations tend to cave easily, because a pile is in essence a stick that is pounded into the ground to develop resistance. Of course, some sticks develop more resistance than others, and that is why there are analysis procedures for designing pile foundations.

Piles are used wherever the near-surface soils are not sufficiently competent to take the proposed structure loads that cause either bearing-capacity failure or excessive settlement and the use of drilled shafts may not be feasible because of the ground conditions. Difficult conditions for piling include hard formations, which resist driving, and boulder deposits, which tend to deflect the piles or break them.

13.2 Types of Piles

There are many types of piles available including timber piles, which are the oldest type of piling. They may or may not be treated chemically to avoid decomposition. This is the type of piling one frequently sees along older wharfs and dock platforms. Their big advantage is that they are inexpensive. However, their big disadvantage is that they are often attacked by marine organisms.

Other types of available piling include steel shapes, steel cylinders or pipes, pre-cast reinforced concrete piles, pre-cast pre-stressed reinforced concrete piles, micro piles (see Section 1810.3.10 in the 2012 IBC), and a multitude of proprietary-type piles such as auger grout injected and concrete pounded into the ground through a pipe. The basic purpose of all piling, no matter what they are called or how they are placed, is to extend the loads to be carried by the soil deeper into the ground. Most piles have a small tip area relative to their side area, and therefore side-skin friction is the predominant method of support for most piles. This would not be the case if the piles could be driven through soft materials to encounter a firm material at depth, in which case the majority of the load would be carried by the pile tip into the firm material.

13.3 Pile Capacity

13.3.1 Design. All the methods of design for piles include distributing the applied load from the structure to the head of the pile, either a single pile or through a pile cap to groups of piles. The load is analyzed as being distributed between tip resistance and skin friction. For long piles in relatively weak soils, tip resistance can be virtually ignored. Skin friction depends on the type of soil material and the depth at which the pile is analyzed. Sand will produce higher resistance as depths become greater because of the increasing confining pressure of the overburden. An "alpha" factor is used for skin-friction design, which is a way for translating the available skin resistance that the pile can experience to the available shear strength of the soil around it. The soil loses something at the interface either due to soil disturbance or other factors. The nature of the skin surface itself is important in the development of skin resistance.

Since piles are generally not as large and do not have the same capacity as drilled pier foundations, they are frequently installed in groups. A pile group is typically attached to a pile cap of reinforced concrete connected across the pile tops. The total resisting force of a pile group is not the same as an individual pile multiplied by the number of piles. This may be the case under certain circumstances; however, normally the pile group acts more like a large irregular-shaped single footing, and it should be evaluated as such. The design for tip resistance and skin friction is similar to that discussed in Chapter 12 for drilled piers.

Design considerations need to be given to the type of material and the protection of the pile material from corrosion or deterioration. Marine borers are a serious problem for wooden pies, and highly acidic soils will attack and destroy steel piles or even concrete piles under certain circumstances.

13.3.2 Test piles (Section 1810.3.3.1.2 of the 2012 IBC). Drilled pier bottoms can be inspected, sampled, and tested if necessary to prove the capacity of the drilled pier. However, with piles there is more uncertainty, and the chief method of determining capacity in the field is by some procedure involving the hammer energy placed into the pile to drive it to a certain depth. The old procedure was to use the "Engineering News Record Formula," which was published in the 1920s and has since been shown to be totally inaccurate. The newer procedures use wave equation analysis, which has to do with the energy put into the pile by the hammer and examination of the return waves to determine the actual holding capacity of a pile. This procedure is not entirely without difficulties, however. There are also indirect methods to determine the integrity of a pile such as the sonic pulse procedure to see if a concrete pile is broken. However, to be certain of the pile capacity on a major job, one or more test piles should be installed, loaded to failure, and the results recorded.

Test piles usually consist of a pair of reaction piles spaced some distance away from the pile to be tested, a spanning beam to transmit the force to the test pile, a jacking ram, a load cell, and dial gauges to measure movement. The procedure has been spelled out in detail in ASTM D 1143. Testing for pull-out resistance is covered in ASTM D 3689. The pile-testing process should always be conducted by knowledgeable people who are specialists in piling technology.

During the pile test, the loads and displacement during the test period are recorded and plotted on a graph, and it becomes readily apparent when the line on the graph bends over that the pile has reached its capacity. Figure 13.1 illustrates a typical pile load-settlement relationship.

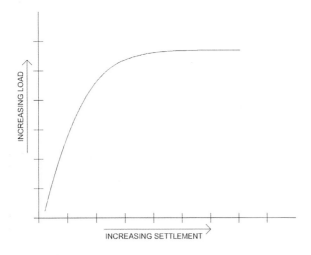

FIGURE 13.1
TEST PILE LOAD AND SETTLEMENT

The data concerning the driving of the test pile, that is, the amount of hammer energy typically expressed in blows per foot of a given type of driving hammer, is also noted on a driving log related to the soil profile that is understood from the test borings. The test-load data will indicate to the engineers whether the effective load-carrying capacity of the pile will be within a design factor of safety. If the value is larger than that assumed in the initial design, construction can proceed. If the resulting load capacity is less than that assumed in design, a redesign is necessary. Often it is important for the engineer to do more than one test pile at various locations on the site because of possible variations in the sub-soil conditions. The blows per foot of the service piles should be compared to the driving data of the test piles for a similar soil profile to verify that each pile has a similar capacity.

13.4   Installation

Basically, the installation of piling is equivalent to pounding sticks into the ground. Of course the sticks can be very large, very long, and of different types. Pile driving includes a supporting crane, the pile, leads for holding the pile steady, a cap cushion, and a driver. Drivers are of many types, ranging from a simple gravity drop hammer to a single- or double-acting steam or compressed-air hammer. Some drivers are hydraulic hammers. A vibratory driver may be used in some cases. See Figures 13.2, 13.3, 13.4, and 13.5 for illustrations of the various types of piling and pile driving equipment in use.

Courtesy of Pile Driving Contractors Association
FIGURE 13.2
TIMBER PILE

Courtesy of Pile Driving Contractors Association
FIGURE 13.3
PRE-CAST CONCRETE PILE

Courtesy of Pile Driving Contractors Association

FIGURE 13.4
PRE-CAST CONCRETE PILING OVER WATER

Courtesy of Pile Driving Contractors Association

FIGURE 13.5
STEEL PIPE PILE

Pile driving should continue until the recorded energy at the head is near what the designer assumed, possibly confirmed by a test pile. Overdriving can damage the pile head or even the tip, and sometimes buckle the entire pile. It is important that the plumb of the pile be evaluated as it is being driven since long piles can deflect and begin to bend upward. They have been known to emerge from the ground some distance away, forming a "U" shape.

The impact vibrations of pile driving may damage nearby structures. This factor should be a consideration during initial project design, and an alternate foundation system may be required.

13.5   TEST QUESTIONS

MULTIPLE CHOICE

1. Name three types of piling generally in use. (select three):
   a. fiberglass
   b. timber
   c. steel pipes
   d. bundled bamboo
   e. pre-cast concrete

2. Name three types of pile drivers (select three):
   a. dead load
   b. gravity drop hammer
   c. double-acting steam hammer
   d. crane pressure
   e. vibratory

3. Pile foundations as a development in construction technology (select one):
   a. were first used 200 years ago to develop the docks of New York
   b. were first used to support cathedrals in the 1400s
   c. were first used to support pre-historic structures thousands of years ago

4. Boulder deposits are difficult conditions for deep foundations; therefore (select one):
   a. piling would normally be used
   b. piling would not normally be used
   c. piling could be used with blasting to break up boulders

5. For a uniform sand deposit, the skin friction along the pile length will (select one):
   a. become greater with depth
   b. be constant
   c. become less due to disturbance
   d. be dependent on the hammer size

6. To determine the capacity of a pile group (select one):
   a. the capacities of individual piles are simply added up
   b. the total capacity of all the piles are multiplied by 50 percent
   c. it is necessary to analyze the group as an irregular footing

7. While applying loads on a test pile versus the measurement of the settlement occurring (select one):
   a. when the point is reached where settlement is increasing without additional load being added, the pile has much more capacity
   b. the maximum capacity is indicated by settlement increasing without adding load
   c. the settlement measured must be multiplied by two

8. Overdriving a pile (select two):
   a. does no harm, since this increases the safety factor
   b. can broom out the top of timber piles
   c. can damage the tip
   d. is a good way to be sure of capacity

9. On a major job, to be certain of the piling capacity, the best thing to do is (select one):
   a. install one or more blue tops
   b. rely on the "Engineering News Record Formula"
   c. trust the opinions of the driving foreman
   d. use test piles

10. To what does ASTM D 1143 refer? (select one):
    a. measuring piles
    b. pile test procedures
    c. strength of concrete
    d. testing reinforcing steel

Chapter 14

RAFT OR MAT FOUNDATION

Heavy Mat Foundation with multiple basements. One Shell Plaza 1968. Note lines of drilled piers and steel raker braces to keep surrounding streets from moving into excavation. In recent years the rakers would likely be replaced with grouted tie-backs, freeing up construction area.

Chapter 14

RAFT OR MAT FOUNDATION

14.1 When to Use Raft Foundations

Rafts, also known as mat foundations, slab-on-ground, or ribbed slabs, are intended to take point or line loads from the superstructure of a building or other facility and spread the loads into a more uniform contact pressure, which eliminates the need for individual foundations and may be quite useful in weak soils. Rafts are shallow foundations since the foundation elements do not penetrate very far beneath the ground surface, and they transfer all the loads to a shallow soil zone. Rafts or mats are also used extensively to overcome the movement of subgrade soils caused by swell or differential settlement. In most applications for small to midsize buildings, the foundation surface is also the floor of the building, eliminating the need for construction of girders, beams, and joists to support the flooring.

Mats are also used for larger buildings and industrial structures in lieu of piles or piers for the same reasons outlined above. In some cases the mat may be several basements below the adjoining ground surface and may be quite thick to produce the requisite stiffness. Mats have been used under buildings as high as 30 to 50 stories, and because of the resulting high column loads that must be distributed to the soil, concrete mats for such buildings can be 6 to 12 feet thick, with extensive reinforcing steel. The use of mats for these large buildings is an economic matter, and the designers compare the cost of driving very deep piles or drilling deep piers in a medium to soft soil to that of constructing a stiffened mat on or near the surface. The effect of pile-driving vibrations on nearby structures or infrastructure may also be a consideration limiting the use of piles. Figure 14.1 illustrates a large mat.

14.2 Types

14.2.1 Stiffened mats.
A stiffened mat is not of uniform thickness throughout, but has thinner slab panels that are bounded by stiffener beams, typically forming a grid structure somewhat like a waffle. Stiffeners can be above the main slab if they coincide with room dividers in a basement, but they are most commonly found below the main slab where they can be formed against earth. Stiffened mats are frequently used in residential construction and mid-size commercial construction in areas where soil conditions warrant such use.

Frequently, these conditions are highly expansive clays, although stiffened mats may also be used on a softer soil that may be subject to settlement or bearing failure. The stiffening beams are placed to provide the necessary stiffness to the mat without using as much concrete as a uniform-thickness mat. Typical residential and light commercial mat foundations with stiffening beams have beam dimensions on the order of 12 inches wide by 18 inches to 48 inches deep, spaced from 10 to 15 feet on centers each way. The stiffening beams are cast monolithically with the floor slab, which can be as thin as 4 inches. The entire assembly of slab and beams acts together to provide a stiffened structure. Figure 14.2 illustrates this construction.

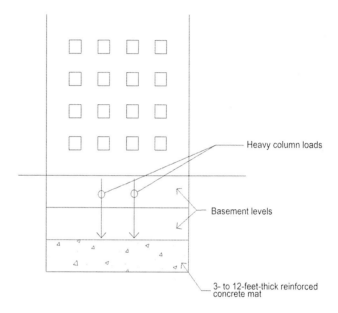

Heavy column loads

Basement levels

3- to 12-feet-thick reinforced concrete mat

FIGURE 14.1
LARGE MAT UNDER A HEAVY BUILDING

14.2.2 Uniform thickness mats.
In addition to typical use in heavily loaded building structures as shown in Figure 14.1, uniform thickness mats ranging from 7 inches to 24 inches thick are often used for residential and light commercial construction. Adequate bending stiffness can be achieved with these mats although the amount of concrete and reinforcing steel is greater than that needed in an equivalent stiffened or ribbed mat. Uniform thickness mats may be used in any situation where a stiffened mat may be used, but the issues are economics and local construction practice. Figure 14.3 illustrates a typical uniform-thickness mat for light construction.

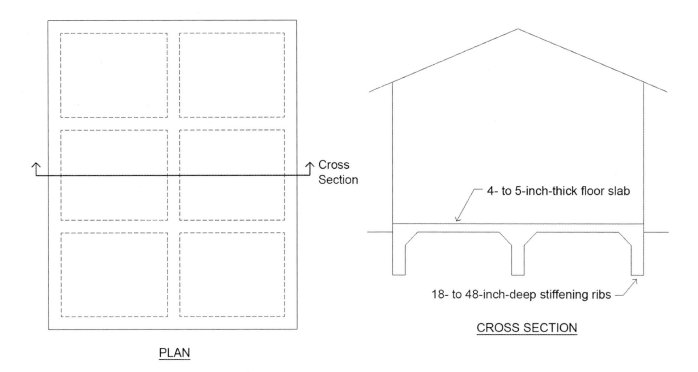

**FIGURE 14.2
TYPICAL PLAN AND CROSS SECTION
OF A STIFFENED MAT**

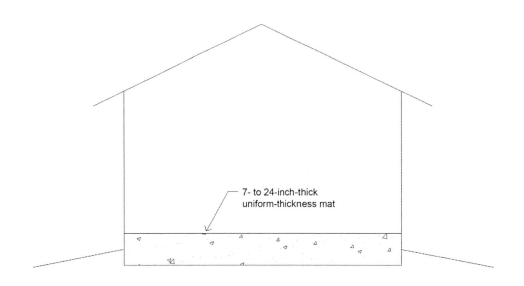

**FIGURE 14.3
CROSS SECTION OF A LIGHT
UNIFORM-THICKNESS MAT**

14.3 Design Considerations

14.3.1 Low-strength soils. If the soils on a site have low bearing capacity, which would require very large spread or strip footings to properly distribute the loads to the soil, the use of a continuous mat, either stiffened or uniform thickness, might be more efficient,. If a crawl space beneath the foundation is not needed, a likely break-over decision point would exist if the area of spread and strip footings exceeds 50 percent of the footprint area of the building. In this case a mat may be applicable and less expensive. The performance of the structure may also be improved because low-strength soils often have varying strengths across the site, and one footing may be in a zone of lower bearing capacity than footings in other areas, leading to differential settlement between footings. A mat will distribute the forces and not permit as much differential deflection across the structure, if properly designed.

14.3.2 Compressible soils. The same considerations for low-strength soils may be applied to compressible soils although the failure due to settlement may not be as dramatic, taking a longer period of time to occur. The distribution of stresses in the soil beneath a mat can vary with the rigidity of the mat as well as the type of soil. For example, a very rigid mat placed on a deep sand deposit will have higher reaction forces in the center of the mat than around the edges. This can be visualized by the sand having a tendency to crawl away around the edges because it does not have as much confinement as that in the middle. Conversely, a very rigid mat placed on a clay soil would tend to have higher stresses around the edges and lower stresses in the middle. This is because the clay acts more like an elastic continuum than sand, and the change in deformation between the outside of the building and the edge of the foundation is greatest around the exterior and less on the interior. See Figures 14.4 and 14.5.

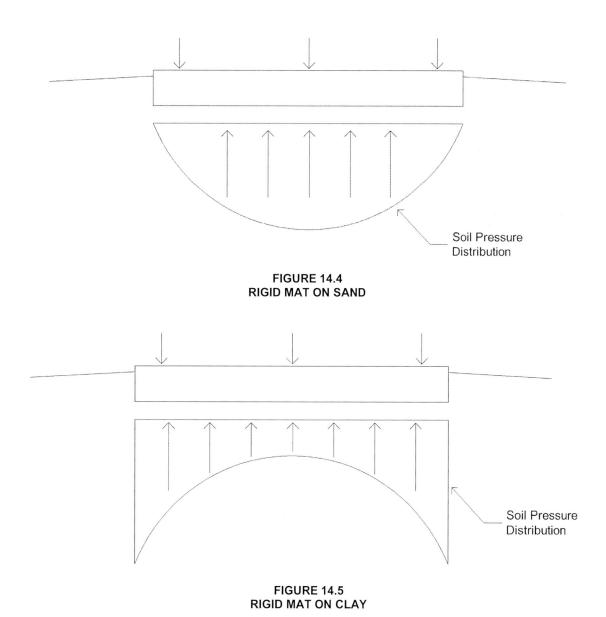

Soil Pressure Distribution

FIGURE 14.4
RIGID MAT ON SAND

Soil Pressure Distribution

FIGURE 14.5
RIGID MAT ON CLAY

On the other hand, a completely flexible mat would tend to not remain plane, but would follow the soil down. In the case of sand, the edges would go down more than the center, and in the case of clay, the center would go down more than the edges. Real mats are somewhere between completely rigid and completely flexible; thus, designers need to take into account the variables above. Another variable is the tolerance of the building itself to distortion, and this will sometimes dictate how much rigidity a mat must have to keep the floor plane and thus keep the building superstructure elements from wracking and cracking.

14.3.3 Expansive (shrink and swell) soils.

For expansive soils, a stiffened or uniform thickness mat is often at the mercy of the soil movements, which can be spontaneous with little regard for the actual pressure placed by the foundation. These types of foundations for residential and light commercial structures usually carry very light building loads. Expansive clay tends to swell or shrink in response the addition or removal of moisture from the soil, primarily due to environmental influences around the edges of the building. Soil-analysis procedures are used to estimate the distance in from the edge that the environmental influences will either dry out the soil causing the clay to shrink away and lose support, or the distance that the moisture can penetrate from the outside during a wet environmental cycle causing the reverse effects. The procedures also define the amount of vertical differential movement between the edge and some point in the interior, approximately where the moisture penetration distances end. These determinations are complex, but usable simplified procedures are described by the Post-Tensioning Institute (PTI), by the Wire Reinforcement Institute (WRI), and in the Building Research Advisory Board Report #33 of the National Academy of Science (BRAB) publications. See Chapter 35 of the 2012 IBC for PTI and WRI references. The BRAB publication may be obtained from the U.S. Government Printing Office. The PTI procedures are more closely based on the principles of unsaturated soil mechanics, while the other methods are empirically based and less detailed. All the procedures require some knowledge of the site soils.

Once a geotechnical engineer makes a characterization of the site soil profile, the resulting support conditions can be modeled to provide a structural solution for these types of mats. In reality this is a very complex structural calculation, and the design procedures that are published, including those of the PTI, WRI, and BRAB, are simplifications to make the complex calculations usable by the typical consultant's office. Figure 14.6 illustrates the support conditions assumed by these procedures.

Residential foundation construction utilizes a shallow foundation of the type described above in many parts of the country. Often, local building officials do not require the design to be engineered, since generally residential structures are not required to have engineered plans. Some jurisdictions will make an effort to identify expansive clays in accordance with Section R403.1.8 of the 2012 IRC. If the site is determined to be an expansive clay site, the IRC refers the reader to Section 1808.6 of the 2012 IBC. This section requires the use of the WRI procedure for conventional rebar slabs, the PTI procedure for post-tensioning reinforcement, or any other procedure that accounts for soil structure interaction, the deformed shape of the soil sup-

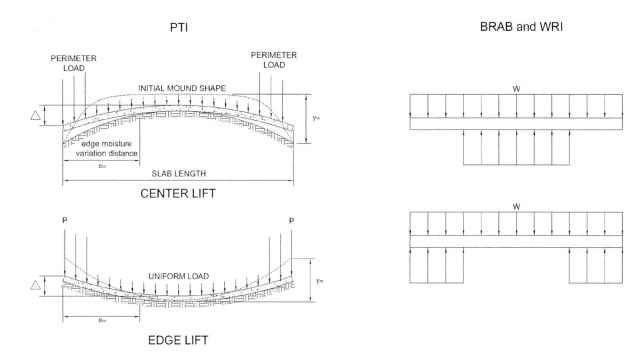

FIGURE 14.6
TYPICAL SOIL SUPPORT CONDITIONS ON EXPANSIVE CLAY FOR A SHALLOW FOUNDATION

port, the plate action of the slab, as well as both center lift and edge lift conditions. BRAB is no longer referenced in the building codes. However, it is still a reliable procedure for rebar slabs if implemented correctly.

14.4 Mild Steel Reinforced Mats

Reinforcement of the concrete in mat foundations, whether the mat is stiffened or of uniform thickness, can be conventional reinforcing steel (rebar) or post-tensioning cables. The mild steel reinforced mats are directly addressed in the WRI procedure and the BRAB procedure. It is also possible to design mild steel reinforced mats using the PTI procedure by zeroing out the effect of post-tensioning cables and supplying minimum ACI 318 steel.

Reinforcement of such mat foundations is based on the need for the addition of strength to maintain the stiffening effect required for the site conditions and to reduce cracking due to concrete drying shrinkage. Mild steel in the slab portion of a stiffened mat with a 4- or 5-inch-thick slab for the purpose of eliminating concrete drying shrinkage typically ranges from #3 bars at 10 inches on center each way to #4 bars at 16 inches on center each way. This, however, is subject to designer decisions. An equivalent amount of welded wire fabric can also be utilized, which should be chair supported as with any other reinforcing steel and not "hooked up" in the wet concrete after the slab is cast. "Hooking" will nearly always result in the reinforcing steel resting on the subgrade soil in the hardened concrete, where it is totally ineffective.

Stiffener beams in a stiffened mat will typically have steel ranging from two #4 bars top and bottom to three #8 bars top and bottom. Figure 14.7 illustrates a typical stiffened mat using conventional steel reinforcement.

Note: In the smaller size bars (up to #8), the diameter of a reinforcing bar goes up by $1/8$-inch increments for each number designation. For example, a #3 bar is 3/8 inch in diameter.

14.5 Post-tensioned Reinforced Mats

Most of the same principles used in mild steel reinforced mats apply to post-tensioning. One difference is that post-tensioning creates a pre-loading of the concrete in compression that bending must overcome before flexural-type cracking can occur. Flexural or bending stresses place the concrete in tension at either the top or bottom face, and the pre-stress force must be overcome before the tensile strength of the concrete is engaged. Flexural cracking of the concrete section greatly reduces the available stiffness. The cables or pre-stress tendons are usually 7-strand, high-strength steel of nominal $1/2$ inch diameter, encased in a plastic sheath. Steel anchors are located on each end of a cable to transfer the pre-stressing force to the concrete. The inherent tensile strength of concrete is typically on the order of 300 psi. Post-tensioning for typical residential and light commercial foundations will add anywhere from 50 to 150 psi to this value, thus increasing the flexural stresses that can be accommodated before cracking.

An advantage of post-tensioned foundations is that the labor required to install the reinforcing cables is considerably less than that required to tie and place reinforcing steel. This is often a factor considered by builders and contractors in selecting this type of foundation. One drawback of post-tensioning reinforcement is that there is an extra step required after the concrete has hardened and the forms are stripped. This step involves stressing the cables with hydraulic rams to a required force level. If this step is not followed, the post-tensioning reinforcement is simply not effective. Figure 14.8 shows a typical layout and section of a ribbed post-tensioned reinforced mat.

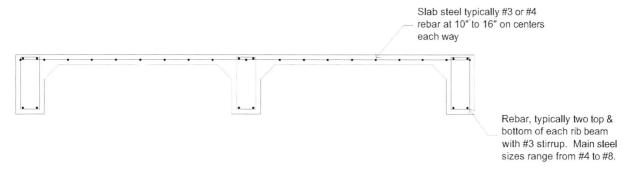

Slab steel typically #3 or #4 rebar at 10" to 16" on centers each way

Rebar, typically two top & bottom of each rib beam with #3 stirrup. Main steel sizes range from #4 to #8.

FIGURE 14.7
CROSS SECTION OF A MILD STEEL REINFORCED MAT FOR A TYPICAL RIBBED (STIFFENED) SLAB

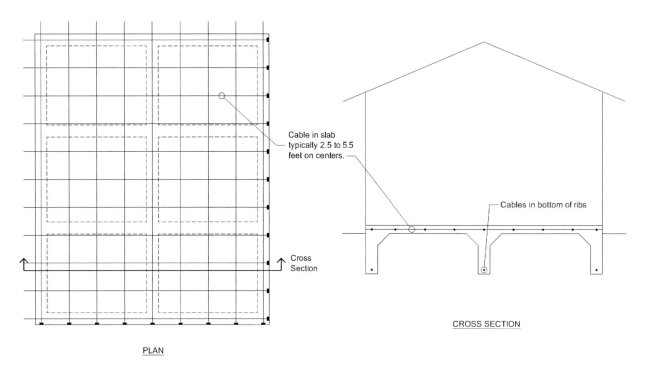

FIGURE 14.8
PLAN AND CROSS SECTION OF A POST-TENSIONED
LIGHT COMMERCIAL OR RESIDENTIAL FOUNDATION

14.6 Light Structures on Good Soils

If the soil is reasonably strong and it does not possess expansive clay properties, a very economical foundation can be established that could consist of shallow stiffener beams around the perimeter, possibly with shallow beams under heavy line-loads on the interior, and a 4-inch flat slab. The slab would be reinforced against concrete drying shrinkage, and the perimeter and interior line-load beams would be typically 12 inches deep. See Figure 14.9 for comparison of typical sections of a stiffened mat on highly expansive soil and a minimal slab on good soil.

14.7 Installation

Since shallow foundations rely on the sub-soil immediately below the foundation for support, it is important that the support be adequate and available in the right places. A stiffened mat foundation, which is often used under residences and light commercial construction, is typically assumed to have its bearing supported on the bottom of the stiffener beams, much like a grid work of strip foundations. A 4- to 5-inch-thick slab panel between the 10 to 15 feet spaced stiffener beams will usually span nominally compacted soil. It is important that the soil beneath the beam bottoms is either intact original soil or properly compacted and tested fill. In many areas the practice is to compact the entire fill pad and test it before trenching the stiffener beams into the soil. This is the more common practice for light commercial construction.

In some parts of the country, residential practice is to extend the beam bottoms into intact soil or compacted and approved fill and to fill the area in between beams to support the slab panels with "forming fill." This fill is only nominally compacted and is intended only to support the concrete until it hardens and can support itself. This only works if the panels are about 10 feet by 15 feet maximum and if the loading on the slab is only that of nominal residential construction. Of course, heavy point loads and line loads must be prepared for differently. In the event that there is considerable fill, and it is not known to be properly compacted, one practice is to carry the perimeter beams through the fill to the intact soil and to carry hard points or interior piers through the fill to support the slab, typically at stiffener beam intersections. Again, the slab panel spacing described above should be followed.

During installation of soil-supported slab foundations, many times it is important to have a vapor-retarder membrane and a capillary-break layer of coarse sand immediately beneath the slab. If moisture rises through the slab, many problems, including wood damp rot and mold as well as separation of floor covering, can occur. If the slab is a stiffened mat, the vapor-retarder membrane has to be installed carefully into the beam trenches so it does not bunch up and interfere with the reinforcement of the beams. Sometimes this is accomplished by cutting the vapor barrier in the bottom of the beam trenches.

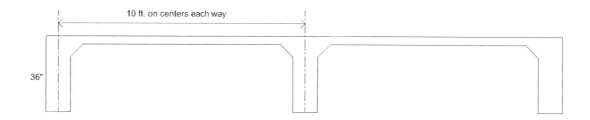

10 ft. on centers each way

36"

STIFFENED MAT ON HIGHLY EXPANSIVE SOIL

12"

NON-STIFFENED MAT ON GOOD SOIL

FIGURE 14.9
RANGE OF TYPICAL SLAB FOUNDATION CONFIGURATION DEPENDING
ON SOIL TYPE. COST DIFFERENCE IS OBVIOUS

Once the pad is prepared, the beams are trenched, and forms are up around the outside (or on the interior as needed for drops), reinforcement is placed, either mild steel or post-tensioning. At this point an inspection is in order to see that things are where they should be, that the steel is chaired up to the proper positions in the beams and the slab, that the vapor retarder is in place, and that any internal drainage features called for by the plans are in place. These drainage features could include back drainage and water-proofing behind drops in slabs, split floor levels, or base-ment walls. After the steel placement and forming is approved, concrete can be placed in the form, filling the beam trenches and the slab in a monolithic placement. Con-crete for these types of foundations is normally specified on the foundation drawings and can range from 2,500 psi to 4,000 psi concrete depending on the requirements of the foundation designer. Observation by an independent inspector of the concrete placement is frequently, but not universally, done. Concrete placement inspection should include observation of placement to avoid displacement of steel or forms, the taking of test cylinders for verification of concrete strength, determination of the slump of the con-crete, determining the air content, and verifying the plant mix tickets.

After the concrete is placed and finished, it is always an excellent idea for the contractor to provide a curing mem-brane, fog spray, or other approved method for keeping concrete moist while the initial strength gain of the con-crete is taking place. For much residential and light com-mercial construction this may not be done, depending on

the local practice and the requirements of the building offi-cial or the design engineer. Curing has the advantage of reducing concrete drying-shrinkage cracking and giving a harder, dust free surface to the concrete. Use of vibration while placing the concrete is also an excellent idea to reduce the amount of "honeycomb" around the forms and to ensure proper consolidation around the reinforcement. Figure 14.10 shows a typical post-tensioned ribbed mat set-up and concrete placement.

FIGURE 14.10
POST-TENSIONED RIBBED MAT SET-UP AND
CONCRETE PLACEMENT

14.8   TEST QUESTIONS

MULTIPLE CHOICE

1. In shrink-swell soils, drying of the supporting soils around the perimeter due to environmental influences will cause the foundation to (select one):
   a. go down in the center
   b. go up along the edges
   c. go down along the edges

2. A uniform-thickness mat, which has an equivalent strength and stiffness to a ribbed mat, is often employed because this type of construction (select one):
   a. uses less concrete and reinforcing steel
   b. uses less hand labor
   c. provides a flatter surface

3. A very rigid mat placed on a deep sand deposit will theoretically have (select one):
   a. much lower reaction forces in the center than around the edges
   b. higher reaction forces in the center
   c. uniform reaction forces

4. In determining how much rigidity or stiffness a mat foundation must have, the following variable should be considered (select one):
   a. the cost of steel
   b. the tolerance of a building to distortion
   c. fracture hinges
   d. the classification of the soil

5. If welded wire fabric is used for reinforcement of a mat (select one):
   a. it does not need to be supported on chairs, but can be hooked into position in the fresh concrete
   b. it must be supported on chairs
   c. it can be of any gage

6. In a post-tensioned foundation (select one):
   a. stressing is not important since the high-strength cable is bonded for its entire length into the concrete
   b. forms are not needed
   c. stressing is important for the cables to be effective

7. The installation of a vapor-retarder membrane and a capillary break layer beneath the slab is (select one):
   a. only important to keep ants out
   b. required to prevent water vapor from penetrating into the building
   c. needed in case of a plumbing break

8. Curing of finished concrete either by a curing membrane or water fog spray will (select two):
   a. reduce concrete drying shrinkage cracking
   b. give a harder dust-free surface to the concrete
   c. is not needed due to cost
   d. is not needed because it is hard to inspect

9. The decision to use a stiffened or uniform-thickness mat as a building foundation may include considerations of whether or not the total of individual spread footing areas exceeds about _____ percent of the footprint area of the building (select one):
   a. 25
   b. 50
   c. 75

10. Slab-on-ground foundations on expansive soil are discussed in 2012 IBC Section (select one):
   a. 1810.1
   b. 1807.2
   c. 1808.6
   d. 1805.1.2

11. What is the diameter of a #4 reinforcing bar? (select one):
   a. 4/10 inch
   b. 1/2 inch
   c. 1/4 inch

Chapter 15

SOIL-SUPPORTED SLAB FLOOR WITH STRUCTURAL FOOTINGS

Typical interior of warehouse with soil-supported slab floor and structural footings

Chapter 15

SOIL-SUPPORTED SLAB FLOOR WITH STRUCTURAL FOOTINGS

15.1 Description

This type of foundation is a hybrid between a soil-supported slab and spread or strip footings or piers that support the main structural loads. Therefore, it could be either a shallow foundation or a deep foundation with a shallow component for the floor slab. This type of construction is frequently used for shopping centers, warehouses, and light manufacturing structures and is sometimes used for residential construction in some parts of the country. For residential construction it works best for non-expansive clay sites. In expansive soil, because the foundation elements are disconnected, a plane surface across the floor of the house is difficult to maintain, and a stiffened or uniform-thickness mat may be more appropriate.

15.2 When to Use a Soil-Supported Slab Floor with Structural Footings

Light commercial or industrial construction may utilize this procedure, sometimes even in expansive soils. In expansive soils the main structural loads, such as the wall and roof frame system, are carried to deep-drilled piers, and the slab is soil-supported with considerations given to the amount of differential heave that may take place because of the clay soils. The heave can be reduced by the upper clays being removed and replaced with an inert fill material or other stabilization techniques discussed in Chapter 16. In many parts of the country good bearing, such as weathered rock, is available only a few feet below the surface, which may or may not have swelling clay above it. This condition would be a good candidate for this type of construction, especially if several feet of the swelling clay can be effectively removed and replaced as described above. In this case the main structural footings could be shallow. This type of construction is frequently used with pre-engineered steel frame or with tilt-wall construction, which are popular procedures for constructing shopping centers and warehouses. Figure 15.1 illustrates a typical hybrid system using shallow spot footings or deep-drilled piers for the main structural loads.

15.3 Design Considerations

It is important for the engineer to consider the use of the structure before designing this type of foundation. Many warehouse loading conditions have high-pressure tire trucks or fork lifts, and the floor slab must be designed as if it were a concrete pavement to avoid being damaged by wheel pressures.

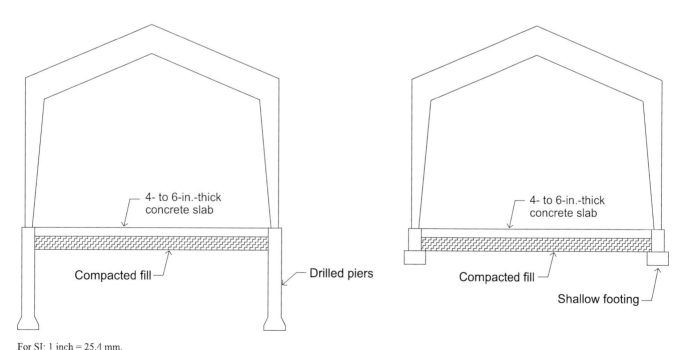

For SI: 1 inch = 25.4 mm.

FIGURE 15.1
TYPICAL HYBRID FOUNDATION

The floor slab is typically placed on a layer of compacted fill. For warehouses or manufacturing buildings this may be 4 to 5 feet thick to provide a truck dock height construction. Because this is a thin, flat slab, it must rely entirely on the fill for support, and the fill should be well compacted and tested for density. It is important that the fill be inert and not consist of expansive clays. Since this type of construction is typically used for large areas, such as shopping centers, warehouses, or industrial complexes, joint spacing for control of slab cracking due to concrete drying shrinkage is important. In addition, isolation joints around pier or footing tops or "plinths" should be supplied with expansion-joint material providing for a slight bit of movement between the more rigidly established footing top and the floor slab, which may tend to move a little bit. Figure 15.2 shows detail of a typical diamond joint that isolates the slab from the major footings; Figure 15.3 is a photograph of this construction detail.

Concrete drying shrinkage requires joint spacing ranging from 12 to 15 feet and is a function both of the shrinkage characteristics of the concrete and the amount of

FIGURE 15.3
DIAMOND BLOCKOUT AT COLUMN FOOTING

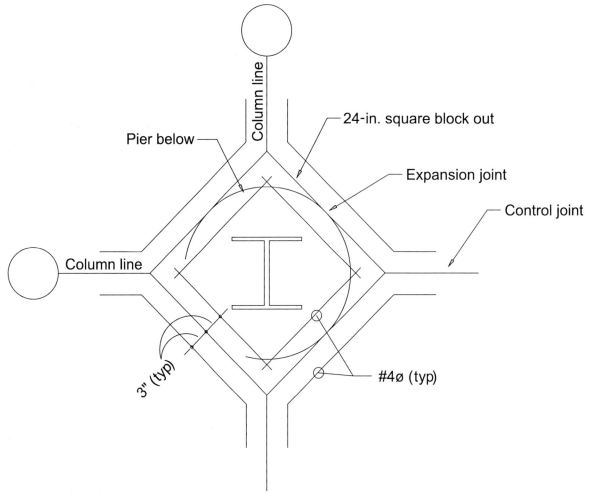

For SI: 1 inch = 25.4 mm, #4 rebar = #13 rebar.

FIGURE 15.2
TYPICAL COLUMN BLOCK-OUT DETAIL WITH EXPANSION JOINTS SEPARATING THE SLAB FROM PIER OR FOOTING.
DIAMOND POINTS CONNECT WITH SLAB CONTROL JOINTS.

reinforcing steel in the slab, which must be determined by the foundation designer. A rule of thumb is that concrete slabs shrink due to drying about $1/8$ inch per 20 horizontal feet. The contraction of the slab is resisted by subgrade friction, setting up tensile stresses. If these stresses exceed the concrete tensile strength aided by reinforcing steel, the concrete cracks.

15.4   Installation

Crack control joints may be tooled in during finishing, formed with a plastic strip during concrete placement, or sawn with a diamond-bladed saw. Joints are formed $1/8$ to $1/4$ inch wide and extend about 10 to 25 percent of the slab thickness. This "controls" the shrinkage cracking to straight lines. Joints are typically filled with an appropriate joint filler. If sawing is used, it should commence very quickly, within 4 to 12 hours of the concrete finishing, or as

soon as the green concrete can support the equipment and sawing will not ravel the concrete. The concrete has already begun to shrink as it is losing water by evaporation, and it is not uncommon in this type construction to see a shrinkage crack occurring one to two inches away from and parallel to the control joint because the joint was sawn too late. This is a tricky problem for contractors and is frequently not done properly. A construction joint is needed where a large slab concrete placement is interrupted, either by design or construction expediency. An expansion joint is one that permits the concrete to expand or contract and is commonly used in concrete pavement and bridge decks where temperature changes are large. Floors inside buildings exist in a controlled environment where temperatures do not fluctuate much, reducing the need for expansion joints. Figure 15.4 illustrates control or construction joint details.

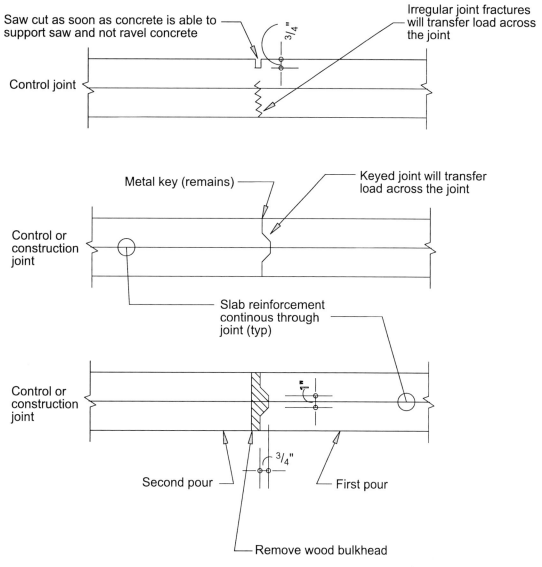

FIGURE 15.4
TYPICAL CONTROL / CONSTRUCTION JOINTS. A CONSTRUCTION JOINT WILL ALSO ACT AS A CONTROL JOINT.

Control of cracking and joint construction is important, especially if the slab is to be the finished floor surface, which may be stained or patterned concrete. Unsightly random shrinkage cracks will greatly detract from the architectural intent as well as provide a fractured area that can enlarge from wheeled traffic. This consideration is frequently overlooked, and it is difficult to find skilled contractors or designers who can produce this work properly. For some industrial or warehouse applications, a hardener may be specified over the concrete surface to resist abrasion from high-pressure tire traffic or movement of heavy, skid-mounted loads. The manufacturer's recommendations should be carefully followed for the use of hardeners.

Curing is important to produce a good concrete surface and reduce drying shrinkage. Thermal expansion and contraction is generally not a problem for these slabs because they are located inside a building with a controlled environment. However, before the building is completed, the slab may suffer considerable thermal shrinkage if, for example, a concrete slab is placed during the day and a drastic drop in temperature occurs overnight. This cooling effect will cause the concrete to shrink, which will be added to the drying shrinkage, resulting in extensive slab cracking.

15.5 TEST QUESTIONS

MULTIPLE CHOICE

1. A type of foundation frequently used for shopping centers, warehouses, and light manufacturing structures could be called a (select one):
   a. warehouse special
   b. hybrid combining a soil-supported slab and footings
   c. slab / footing floor

2. For this type of construction, the floor is a flat non-stiffened slab resting on fill. The inspection of this fill and compaction control (select one):
   a. is not important
   b. is very important
   c. can be done by the "heel test"

3. Drying shrinkage crack-control joints should be installed by the contractor after finishing the concrete (select one):
   a. at any time within two weeks
   b. within 4 to 12 hours
   c. during the next overcast day
   d. at any convenient spacing

4. In warehouse and light manufacturing facilities, loading conditions on a slab may include high-pressure tire trucks or fork lifts, and the floor slab must be designed as if it were a (select one):
   a. skid surface
   b. pavement
   c. reservoir floor

5. Concrete drying shrinkage requires control-joint spacing in these types of slabs ranging from (select one):
   a. 3 to 6 feet
   b. 20 to 30 feet
   c. 12 to 15 feet

6. A "rule of thumb" is that concrete slabs shrink due to drying about _____ inches per 20 feet horizontal distance (select one):
   a. 2
   b. 3/4
   c. 1/8
   d. 1

7. A concrete floor slab will be subjected to drying shrinkage possibly leading to unsightly cracking. What other factor may cause excessive shrinking? (select one):
   a. rain
   b. expansive cement
   c. rapid cooling
   d. downhill creep

8. If skid loads and hard forklift tires are expected to be present, what should be considered? (select one):
   a. higher subgrade compaction
   b. broom finish
   c. surface hardener

Chapter 16

SITE-STABILIZATION TECHNIQUES

Water injection to stabilize expansive clay site

Chapter 16

SITE-STABILIZATION TECHNIQUES

16.1 Reasons for Site Stabilization

If the near-surface soils are unstable and generally unsuitable for foundation construction, stabilization may be an option. Unstable soils may be too soft and weak to properly support the foundations without excessive settling or foundation failure. Also, there could be sites that have active expansive clays, which could be very troublesome, especially if light structures on shallow foundations were constructed on them. Stabilization is also used in subgrades for roads and other pavements. There are several options that can be used for site stabilization, and all will hinge around a cost-effectiveness analysis. The depths and types of soil stabilization should be determined and recommended by the geotechnical engineer.

16.2 Remove and Replace Soil

16.2.1 Use of moisture-conditioned soil. If the soil is unstable because it is too expansive, it may be excavated to a sufficient depth to relieve the problem at the surface of the ground. The material that is removed can be replaced with the same soil that came out of the excavation, provided it is not organic or trashy, by moisture-conditioning it. For example, expansive clay is usually hard, relatively dry clay with moisture content near the Plastic Limit. Excavated soil of this type that is mixed with sufficient moisture to bring the moisture content above the Plastic Limit and then replaced in the excavation may alleviate the expansive problem. Construction specifications typically call for field moisture during compaction to be 1 percent to 5 percent above the Proctor optimum moisture. Control of the moisture during construction with non-uniform soils may be difficult because different soils will have different Proctor optimum moisture contents.

16.2.2 Replacement with select fill. Replacement with select fill is an option frequently employed with either soft unstable soils or expansive clays. The offending soil is removed to an appropriate depth, and a select fill is used as a replacement. The select fill may be silty sand, gravel, crushed stone, or local material treated with a chemical such as lime to render it more stable and compactable. The depths of removal and replacement with select fill will vary depending on the problem and the effect that can be calculated by the geotechnical engineer. Naturally, greater depths of removal and replacement are more expensive than removal and replacement at shallower depths. The typical replacement for an expansive clay site ranges from 2 to 10 feet. If the depths become too great, the use of a different type of foundation, such as a structural suspended pier and beam system, may be more feasible.

Wet and soft unstable soils will need to be removed to the depth at which their adverse effect will not be a problem to shallow foundations. It is usually not feasible to attempt to dry out wet soil and replace it, since the cause of the original wetness may return, and the soil may again be saturated. In the case of highly organic soil or peat, nearly all of the material must be removed to avoid unacceptable settlements.

Figure 9.7 is a cross-section view of an expansive site with select fill replacement. If select fill replacement is used, the excavation should be provided with a subdrain system so water does not accumulate in the bottom of the select fill, which is generally more coarse and pervious than the clay soils removed. Accumulated water will generally increase the swelling clay problem, causing more harm than good.

16.3 Chemical and Water Injection

Often it is appropriate for chemical fluids to be injected to modify the soil. Plain water may be injected to pre-swell an expansive clay site. The depths of injection typically range from 5 to 10 feet. The chemicals used in this process may consist of lime slurry or a proprietary chemical product. The purveyors of the proprietary products are somewhat secretive about the composition of their products, and if this material is used, adequate information should be provided by the manufacturer demonstrating its effectiveness, including laboratory tests and full-scale, long-term observations of its use. The effect of the chemicals, whether they are lime or something else, is to alter the mineralogical composition of clay soils to reduce their tendency to expand when water is made available to them. Water injection simply pre-swells the soil, and frequently chemical injection is another form of water injection because water comprises over 90 percent of the product injected into the ground.

In all cases when chemical or water injection is used to stabilize a site, sampling should be performed on fairly close spacing after the process is complete to ensure uniform distribution of the injected material and to test the actual properties of the soil to verify the desired effect. Some results can be oversold by the purveyors of the product or process, and the promises offered concerning the results of using these products should be taken with a grain of salt. The products and procedures do have desirable effects in many cases, but it is important to verify some sound basis for their use. The 2012 IBC sections 1808.6.3 and 1808.6.4 discuss removal and replacement and stabilization techniques in expansive soil.

Figure 16.1 illustrates a schematic of chemical or water injection into the soil by using a ganged set of injectors, usually mounted on a tractor. Spacing of such injections ranges from $2\frac{1}{2}$ to 5 feet on centers and may be repeated if the injections are not totally satisfactory based on subsequent sampling and testing.

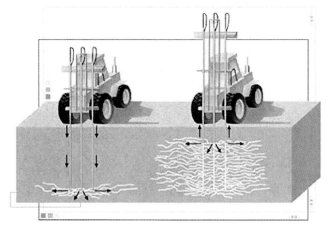

Courtesy of Hayward Baker

FIGURE 16.1
LIME INJECTION FOR EXPANSIVE SITE STABILIZATION

16.4 Vertical Foundation Moisture Barriers

Differential movement of a shallow foundation in expansive clays is caused by penetration of moisture under the slab from around the perimeter (causing swelling) or loss of moisture by evaporation from under the slab around the perimeter (causing shrinking). If this moisture change can be stopped or greatly slowed, the harmful effects of differential moisture change and the resulting soil volume change can be avoided. A vertical moisture barrier has been used in many cases extending to depths of 3 to 12 feet depending on geotechnical analysis and recommendations. If such a barrier is used, it should be designed and detailed with care and constructed adequately with detailed inspections since it is quite easy for these installations to lose their effectiveness through gaps or tears that go unnoticed. The best procedure is to use a heavyweight flexible membrane, which is very difficult to tear with a shovel or pick point, and adhere it tightly to the side of the foundation perimeter. Sometimes a subsurface drain is combined with a barrier on the outside of the barrier to remove any excess ground water by gravity. Figure 16.2 illustrates a typical vertical foundation moisture barrier.

Figure 5.3 in Chapter 5 illustrates a vertical moisture barrier with a subsurface drain in the case of excessive groundwater, which may appear seasonally or constantly on an expansive clay site.

16.5 Horizontal Foundation Moisture Barriers

Horizontal barriers have approximately the same effect as vertical foundation barriers. They also must be well connected to the foundation perimeter and be placed below the ground to a sufficient depth to avoid being disturbed. A danger of these type barriers is that someone will come along later and while digging holes to plant trees or cut utility trenches, compromise the barrier. Therefore, it is a good idea to put a 4-inch layer of concrete over the horizontal barrier under the ground to at least warn people that there is something important underneath. Horizontal barriers typically extend to 10 feet outward. An effective barrier may be composed of pavements or concrete flatwork, such as sidewalks, provided the juncture with the foundation and the joints are properly sealed against water intrusion. Water injection, chemical injection, and moisture-conditioned soil may be made more effective with a horizontal barrier that prevents a future loss of the added moisture around the perimeter due to evaporation.

16.6 TEST QUESTIONS

MULTIPLE CHOICE

1. Name three techniques that can be used for soil stabilization (select three):
   a. vertical moisture barriers
   b. grass mats
   c. chemical injection
   d. burying stumps
   e. remove and replace

2. Stabilization of near-surface soils under a foundation may be an option to be considered (select two):
   a. if the soils are unstable
   b. if the soils are expansive
   c. if rain is predicted
   d. if the building floor is subject to wheel loads

3. The intended effect of injected chemicals in expansive clay soils is to (select one):
   a. eliminate termites
   b. alter the mineralogical composition of the soil
   c. remove the need for controlled compaction

4. An option frequently employed with either soft, unstable soils or expansive clays would be removal and replacement with (select two):
   a. CLSM
   b. timber grillage
   c. select fill
   d. moisture conditioned soil

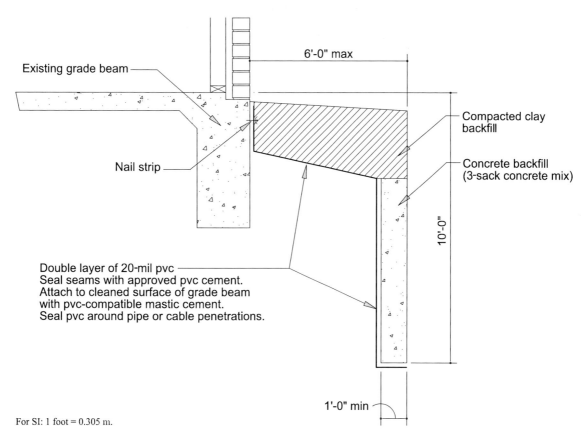

Existing grade beam

6'-0" max

Compacted clay
backfill

Concrete backfill
(3-sack concrete mix)

Nail strip

10'-0"

Double layer of 20-mil pvc
Seal seams with approved pvc cement.
Attach to cleaned surface of grade beam
with pvc-compatible mastic cement.
Seal pvc around pipe or cable penetrations.

1'-0" min

For SI: 1 foot = 0.305 m.

FIGURE 16.2
TYPICAL VERTICAL FOUNDATION MOISTURE BARRIER FOR USE IN EXPANSIVE CLAY

5. The typical removal and replacement depth for an expansive clay site may range from (select one):
   a. 15 to 20 feet
   b. 6 to 10 inches
   c. 2 to 10 feet
   d. 3 to 5 feet

6. The depths of injection of water or chemical products typically range from (select one):
   a. 5 to 10 feet
   b. 1 to 3 feet
   c. 15 to 20 feet

7. Injections of water into an expansive clay soil will (select one):
   a. create a muddy and unstable site
   b. cause an unstoppable swell
   c. pre-swell the soil

8. Water or chemical injection usually employs pressure injectors with spacing of (select one):
   a. 1 to 2 feet
   b. 5 to 10 feet
   c. 2.5 to 5 feet

9. Name two types of foundation moisture barriers (select two):
   a. poured in place
   b. horizontal
   c. sheet piling
   d. vertical

10. Vertical moisture barriers typically range in depth from (select one):
   a. 1 to 2 feet
   b. 3 to 12 feet
   c. 10 to 30 feet

Chapter 17
RETAINING STRUCTURES

Failure of retaining wall

Chapter 17

RETAINING STRUCTURES

(SECTION R404 OF THE 2012 IRC AND SECTIONS 1610, 1806, AND 1807.2 OF THE 2012 IBC)

17.1 When Retaining Structures Are Needed

When the earth is cut or will be filled to approximate a vertical condition, some type of retaining structure is needed. The only exception may be in a cut of sound rock, which will retain itself quite well on a vertical face in many cases. An earth cut will attempt to reestablish its "comfortable" slope, and a retaining wall is necessary to keep it in position.

17.2 Types of Retaining Structures

Retaining structures are generally of two types, gravity structures and cantilever structures. Within these two general types many variants are possible. A gravity structure employs a crib system, a depth of stacked rock, or soil anchorage to produce a block of material that is stable in its own right, sufficient to resist the lateral forces of the soil behind it without sliding or overturning.

Cantilever structures are usually reinforced concrete and employ a foundation that is securely supported by the ground and a vertical element, which is structurally capable of resisting the horizontal force of the soil. These are probably the most of common types of retaining structures in use, and concrete cantilever retaining walls are frequently seen.

There are also hybrid retaining structures that could be compound cantilever structures, such as cantilever walls with tiebacks to "dead men" or anchor piers further back from the front of the walls. Anchored bulkheads typically are sheet piling with an anchored tieback.

Figures 17.1 thru 17.6 illustrate various types of retaining structures.

17.3 Various Limitations

Almost all of the retaining structure types discussed above are subject to some lateral movement before they reach stability. This movement may be very small, on the order of fractions of an inch. However, if other structures are to be supported on the soil above this wall or connected to it, such movement must be taken into account. Least likely to exhibit lateral movement are bridge abutments, basement walls, and split-level structure walls, which may have only a very small amount of movement and could be considered rigid. The various gravity structures and the stacked pre-cast block walls all tend to move somewhat, and designs for pavements or building foundations above them should consider this. The selection of a retaining structure is based on economics and whether or not a vertical face is required versus a sloped face.

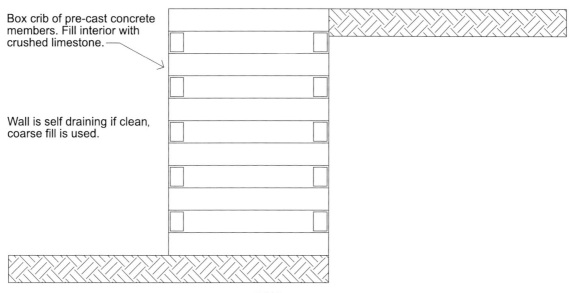

Box crib of pre-cast concrete members. Fill interior with crushed limestone.

Wall is self draining if clean, coarse fill is used.

FIGURE 17.1
CRIB GRAVITY WALL

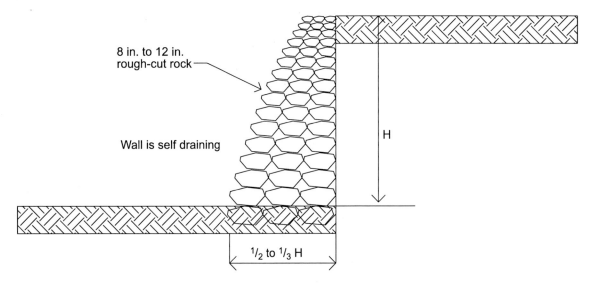

8 in. to 12 in.
rough-cut rock

Wall is self draining

H

$^1/_2$ to $^1/_3$ H

**FIGURE 17.2
STACKED ROCK GRAVITY WALL**

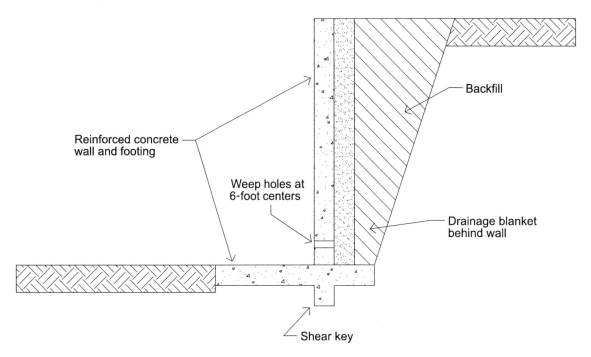

Reinforced concrete
wall and footing

Weep holes at
6-foot centers

Backfill

Drainage blanket
behind wall

Shear key

**FIGURE 17.3
CANTILEVER CONCRETE WALL**

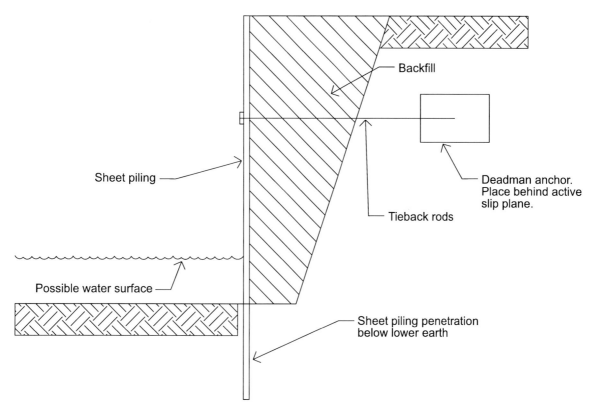

FIGURE 17.4
ANCHORED BULKHEAD (OFTEN USED AT WATERFRONT)

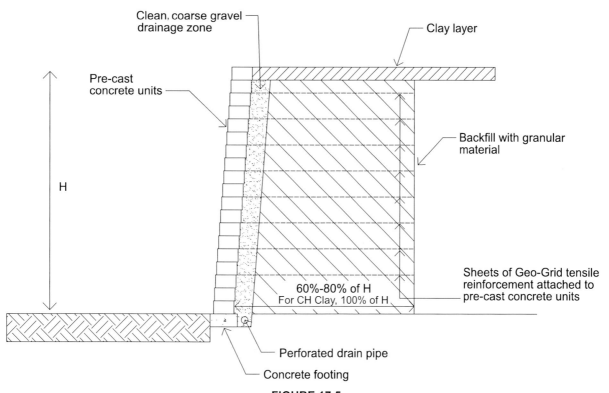

FIGURE 17.5
SEGMENTAL CONCRETE WALL WITH TENSILE ANCHORAGE.
THESE WALLS ARE ALSO KNOWN AS MECHANICALLY STABILIZED EARTH STRUCTURES (MSE).

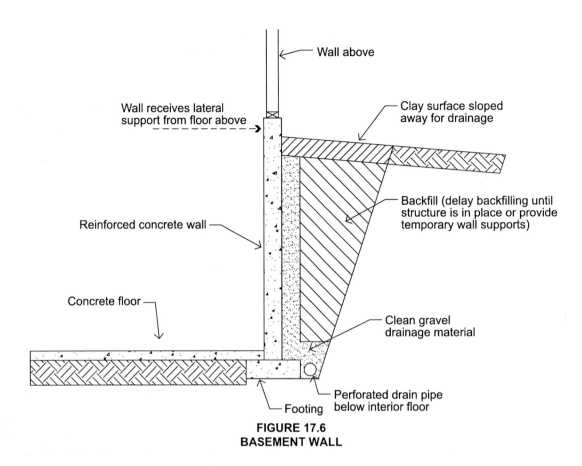

FIGURE 17.6
BASEMENT WALL

17.4   General Design Principles

17.4.1 Earth pressures

As discussed in 8.7, the term "earth pressures" refers to the horizontal forces on a retaining structure generated by the retention of the earth.. The magnitude of such pressures is determined by the type of soil material, the height of the soil material, drainage conditions, and the rigidity of the wall itself.

Design of major retaining structures can be a complex topic, and this book is not intended to cover this category of wall in detail. A qualified geotechnical engineer and structural engineer should collaborate in this type of design.

Horizontal earth pressures are typically assumed to be a triangular-shaped loading, which means that the horizontal pressures near the bottom of the back of the wall are much greater than those near the top. The soil is considered to be an equivalent fluid ranging from a unit weight of about 30 pcf to over 100 pcf. The heavier the unit weight and the greater the height, the more the lateral pressures; see the discussion in 8.7. In triangular loading distribution it is assumed that the horizontal forces are totaled and concentrated at a point one-third of the distance up from the base

of the wall, and this loading concentration is used by the designer to determine the overturning resistance of the wall as a whole and to design the elements of the wall structurally to avoid being damaged by the forces.

Figure 17.7 illustrates the triangular lateral-force distribution behind a retaining wall. With this force distribution available, the structural designer can then analyze the wall for general stability as well as the ability of the wall to keep from breaking off under the bending forces applied to it.

Table 17.1 is Table 1610.1 from the 2012 IBC showing equivalent fluid pressures for lateral soil loadings; this is discussed in more detail in 8.7. In this table, the design lateral soil loads are given in psf per foot of depth, which conveniently works out into pcf of equivalent fluid as discussed in Chapter 8. This table is for backfill against basement or retaining structures, but no soil loads are given for the classifications of OL, MH, CH, and OH. Sometimes these materials require a retaining-wall design: see Table 8.2, Types 4 and 5. The OL, MH, and OH classifications should be designed for the unit weight of water plus the soil material submerged unit weight typically using a total of 100 pcf for both active and at-rest pressures. CH clays should use 120 pcf.

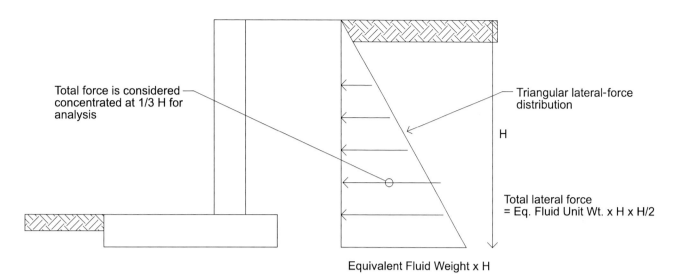

Total force is considered concentrated at 1/3 H for analysis

Triangular lateral-force distribution

H

Total lateral force = Eq. Fluid Unit Wt. x H x H/2

Equivalent Fluid Weight x H

FIGURE 17.7
TRIANGULAR FORCE DISTRIBUTION AGAINST A RETAINING WALL. ASSUMPTION TYPICALLY USED IN DESIGN

TABLE 17.1
SOIL LATERAL LOAD (TAKEN FROM 2012 IBC, TABLE 1610.1)

DESCRIPTION OF BACKFILL MATERIAL[c]	UNIFIED SOIL CLASSIFICATION	DESIGN LATERAL SOIL LOAD[A] (POUNDS PER SQUARE FOOT PER FOOT OF DEPTH)	
		ACTIVE PRESSURE	AT-REST PRESSURE
Well-graded, clean gravels; gravel-sand mixes	GW	30	60
Poorly graded clean gravels; gravel-sand mixes	GP	30	60
Silty gravels, poorly graded gravel-sand mixes	GM	40	60
Clayey gravels, poorly graded gravel-and-clay mixes	GC	45	60
Well-graded, clean sands; gravelly sand mixes	SW	30	60
Poorly graded clean sands; sand-gravel mixes	SP	30	60
Silty sands, poorly graded sand-silt mixes	SM	45	60
Sand-silt clay mix with plastic fines	SM-SC	45	100
Clayey sands, poorly graded sand-clay mixes	SC	60	100
Inorganic silts and clayey silts	ML	45	100
Mixture of inorganic silt and clay	ML-CL	60	100
Inorganic clays of low to medium plasticity	CL	60	100
Organic silts and silt clays, low plasticity	OL	Note b	Note b
Inorganic clayey silts, elastic silts	MH	Note b	Note b
Inorganic clays of high plasticity	CH	Note b	Note b
Organic clays and silty clays	OH	Note b	Note b

For SI: 1 pound per square foot per foot of depth = 0.157 kPa/m, 1 foot = 304.8 mm.

a. Design lateral soil loads are given for moist conditions for the specified soils at their optimum densities. Actual field conditions shall govern. Submerged or saturated soil pressures shall include the weight of the buoyant soil plus the hydrostatic loads.

b. Unsuitable as backfill material.

c. The definition and classification of soil materials shall be in accordance with ASTM D 2487.

Section 1806 of the 2012 IBC, Section R404.5 of the 2012 IRC, and good practice require a factor of safety of 1.5 against sliding or overturning of retaining walls.

17.4.2 Bearing and sliding resistance.

In addition to the horizontal earth pressures applied to a wall, the resistance to bearing failure under the wall footing and the resistance to sliding of the entire structure along the surface of the soil must be evaluated by the designer. In the case of a cantilever retaining wall, the horizontal footing is being pushed away from the upper soil tending to cause overturning, and the soil pressure is increased at the front toe of the wall footing compared to the rear or heel. A check should be made to be sure that the front soil pressures do not exceed the safe bearing pressures of the soil at that location.

Sliding resistance is estimated based on the friction factor between the base of the wall and the ground, for both a gravity wall and cantilever concrete wall, to restrain the full force of the horizontal pressures. This resistance is usually expressed in terms related to the vertical load or "normal" (vertical) pressure on the footing multiplied by a friction factor. The friction factor must be established by the geotechnical engineer, and the normal forces can be calculated by the structural engineer. Clay subgrades commonly are considered to provide "cohesion," not dependent on the normal force. Some designs utilize a resistance at the front of the footing, which is buried in the soil a few feet to aid the sliding resistance. This is not always a good idea since the front edge may be removed by erosion or excavation and may not be present when needed. A key is often used under cantilever walls to increase the sliding resistance.

Table 1806.2 of the 2012 IBC provides guidance for lateral bearing and sliding resistance. Figure 17.8 illustrates sliding resistance and increased bearing pressures under the toe of a cantilever concrete wall.

17.4.3 Active and at-rest pressures.

A bridge abutment, a basement wall, or split-level structure wall will be virtually rigid when subjected to horizontal forces. In these cases the horizontal pressure values used for design will be greater than if the wall were permitted a slight amount of movement because if the soil is permitted to move a little bit, the small strains in the soil mass increases resistance to further movement and reduces the amount that has to be added by the resistance of the wall. If the small amount of movement is not permitted because the wall is rigid, the pressures will be higher. If the wall type permits a small amount of movement of a fraction of a percent of the wall height, this would be analyzed as an "active pressure" case. "At-rest pressures" applicable to rigid walls can nearly double the horizontal pressures on a rigid retaining structure and must be considered in the design.

17.4.4 Reinforcement steel in concrete walls.

Reinforcement steel in a cantilever concrete wall is subject to high tensile forces, which are maximized at the base of the wall adjacent to the earth face. Therefore, the steel must be embedded in the concrete footing and splice-lapped with the wall steel as necessary to provide a continuity of steel in this area. It is also important that the concrete have adequate cover of at least 3 inches of concrete on the earth side to avoid corrosion of the steel. Many concrete walls have failed after 10 or 20 years of service because the steel has been rusted through and lost its capacity to resist the tensile forces applied to it.

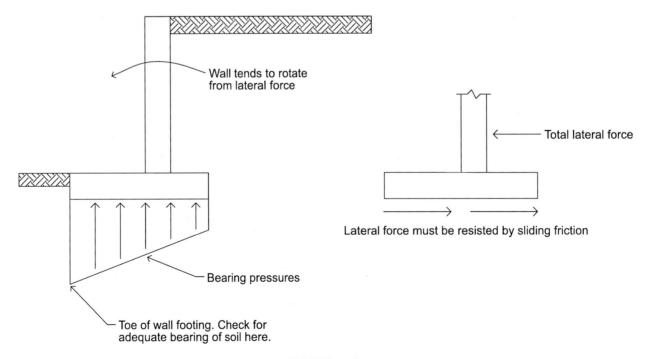

Wall tends to rotate from lateral force

Bearing pressures

Toe of wall footing. Check for adequate bearing of soil here.

Total lateral force

Lateral force must be resisted by sliding friction

FIGURE 17.8
INCREASED FOOTING TOE PRESSURE DUE TO OVERTURNING LATERAL FORCE AND SLIDING RESISTANCE OF WALL FOOTING

The wall of a cantilever concrete retaining wall will have reinforcement steel running vertically along the earth side, usually diminishing in size or increasing in spacing toward the top because the forces are less toward the top. The footing will also be reinforced because it will be subject to bending as the wall tries to turn over. In addition to this steel, secondary steel to guard against shrinkage or thermal cracking is placed perpendicular to the primary or main steel. Provisions for the placement and size of steel are covered by the American Concrete Institute codes, and sizing and placement are the wall designer's responsibility. It is also important that the steel in these walls be clean so it can bond to the concrete properly. The steel should not be covered with grease or dirt, which would cause it to lose connection to the concrete. Concrete retaining walls should have a full vertical expansion joint with slip steel dowels at 100-feet spacing to allow for thermal change. The horizontal reinforcing steel should not go through this type of joint. Vertical contraction or shrinkage-control joints should be provided at 20-feet intervals. These are created by a triangular-shaped strip on the inside of the outside form. The horizontal deformed reinforcing bars should run through this type of joint. Section R404 of the 2012 IRC and Section 1807 of the 2012 IBC provide guidelines for design of concrete and masonry walls. Figure 17.9 shows typical retaining wall reinforcing steel.

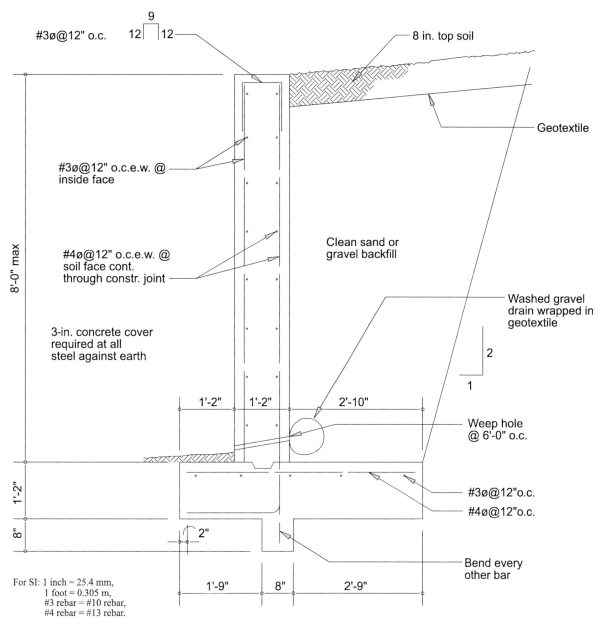

FIGURE 17.9
TYPICAL RETAINING WALL REINFORCEMENT

17.5 Water and Retaining Structures

The majority of retaining-structures failures are caused by water. If a well-drained retaining structure does not build up a head of water behind it, the design parameters described previously will generally be valid. If, however, water is allowed to build up behind the wall, it is necessary to add 62.4 pounds per cubic foot of unit weight, which is all translated into horizontal pressure added to the buoyant soil equivalent fluid design numbers. Thus, buildup of water behind a retaining wall will double or more than double the horizontal forces.. This can readily lead to a loss of the factor of safety and the failure of the wall.

For this reason the walls must be properly drained, either through the wall or by a drainage system behind the wall that will carry any water build up away before it accumulates. The anchored steel bulkheads, crib walls, stacked-rock gravity walls, and segmented concrete walls with tensile anchorage are generally self-draining through the openings in the rock or steel faces of the walls. However, it is still a good idea for the designer to indicate a clean granular drainage material behind these walls, usually with a geotextile separator to avoid the loss of sand or clogging by silt. Cantilever concrete walls may be drained through the face of the wall near the base by use of weep holes, which are typically 4 to 6 feet on centers and about 6 inches above the lower grade. These weep holes are typically 1 to 2 inches in diameter, should penetrate the wall, and should contact a drainage material behind the wall, such as a clean sand or washed gravel, to relieve any water that builds up behind the wall.

If the wall is facing an occupied area or other area at which water is not desirable, such as patios or sidewalks, it may be necessary to place a continuous drain behind the wall using a granular drainage material with a perforated pipe sloped to run out at some point to daylight. The pipe should slope continuously at a minimum slope of 0.5 percent and should be a minimum of 6 inches below the floor slab. The construction of this drainage feature is frequently overlooked or improperly done during construction, and water will build up behind the wall and exit into the finished spaces causing a great deal of occupant irritation or actually causing the failure of the wall.

Figure 5.4 in Chapter 5 illustrates drainage behind a wall that is enclosing a finished space, such as a basement wall or a split-level construction.

17.6 TEST QUESTIONS

MULTIPLE CHOICE

1. 2012 IBC Table 1610.1 lists design lateral soil loadings. In this table is a footnote which advises that (select one):
   a. submerged or saturated soil pressures cannot be ignored
   b. hydrostatic loads can be ignored
   c. soils must be kept moist

2. Cantilever retaining wall soil pressures from the footing are (select one):
   a. greatest at the rear
   b. greatest at the front toe of the wall footing
   c. uniform across the footing

3. Water build-up behind retaining structures (select one):
   a. always happens
   b. is not normally a problem
   c. can lead to wall failure

4. Weep holes (select one):
   a. are typically placed near the bottom of a retaining wall if it is facing an occupied area
   b. cannot be used at an occupied area
   c. are never used

5. When the earth is cut or will be filled to approximate a vertical condition (select one):
   a. the cut face will stand indefinitely depending on soil type
   b. some type of retaining structure is needed
   c. rip-rap is needed

6. What are the two main types of retaining structures? (select two):
   a. tie backs
   b. gravity
   c. braced
   d. cantilever

7. A crib system is what type of retaining structure? (select one):
   a. box
   b. open
   c. gravity
   d. supported rock

8. The most common type of retaining structures in use are (select one):
   a. sheet piles
   b. cantilever concrete retaining walls
   c. hybrid
   d. gunite

9. A segmental concrete wall with tensile anchorage generally employs sheets of geo-grid tensile reinforcement extending back to equal _____ percent to _____ percent of wall height (select one):
   a. 10 to 20
   b. 40 to 50
   c. 60 to 100
   d. 50 to 55

10. In retaining structure design, horizontal earth pressures are typically assumed to be (select one):
    a. rectangular-shaped loading
    b. polygon-shaped loading
    c. triangular-shaped loading
    d. rhomboid-shaped loading

11. In determining horizontal pressures on retaining structures, the engineer considers the soil to be an equivalent fluid with unit weights of about (select one):
    a. 10 pcf to over 20 pcf
    b. 30 pcf to over 100 pcf
    c. 25 pcf to over 45 pcf

Chapter 18
SLOPE STABILITY

Typical small land slip

Chapter 18

SLOPE STABILITY

18.1 Types of Slope Instability

Slope instability can be broadly categorized into three types: toppling (falling) of rock or hard-cemented soils, much like a glacier "calving" into the sea; a gradual downhill creep; and an existing or potential landslide.

Toppling failures may be seen on vertical rock bluffs, often at the side of oceans or lakes, where water has softened the supporting geologic formation below the base of the rock. As the formation below is softened, the weight of the rock above compresses and squeezes out the formation below, and the rock loses its support, generally resulting in vertical splits. When it leans sufficiently, it falls. Toppling failures also include boulders coming loose from a very steep rock cut, sometimes into people's yards, people's houses, or highway right-of-ways. Toppling failures can also be seen along sea shores where wave action erodes the base of cliffs.

Soil sites with slopes more than 6 percent may be subject to either downhill creep or total landslide. The instability could range from gradual downhill creep, which is difficult to detect, to an existing or potential landslide, which could be devastating to the project. Saturated soils on a slope are always suspect and may fail catastrophically, especially during seismic events.

18.2 Dangers of Slope Instability

Instability of a slope can result in gradual movement over a long period of time, which will damage any constructed facilities above or on the slope. If a sudden catastrophic failure occurs, commonly called a landslide, a devastating loss of the structure on the slope, below the slope, or even on top of such a slope could occur, possibly including loss of life. In any event, the failure is generally significant and very difficult and expensive to remediate. Figure 11.2 illustrates the IBC and IRC provisions regarding slopes and structures.

18.3 Stabilization Techniques

Various procedures are available to stabilize an unstable slope. They include drainage, slope reduction or flattening, structural solutions, or chemical stabilization. Stabilization of an unstable slope will require the services of experts and specialists in this type of work, and it can be very difficult to determine the cause of the instability or to arrive at the best solution for stabilization.

18.3.1 Sub-drainage. Often a slope becomes unstable because of seepage water, which may occur constantly or intermittently depending on rainfall history. Because seepage water softens the soil, causing it to lose strength, and at the same time causes the affected soil to gain weight because of saturation, sub-drainage can frequently be effective in restoring a satisfactory factor of safety to such slopes. The engineer of the drainage remediation will have to take into account the depth to which the water must be removed; then a subsurface drain using porous media and gravity outfall pipes may be employed. Occasionally, simply diverting surface water at the upper part of the slope can be effective in reducing the saturation of a slope.

18.3.2 Slope reduction. On a soil slope beginning with a vertical cut, the slope at vertical is most unstable, and frequently it fails even before the cut is complete. As the face of the slope is flattened toward horizontal, the driving forces on the slope are reduced proportionately. There is some angle of the face of the slope at which the slope will remain stable. Nature frequently finds these stable configurations over time by repeatedly failing slopes until they are flattened to the point where the site conditions and the soil strength are in equilibrium. Removing the upper portions or head of a slope or a landslide is a frequently employed technique for reducing the driving forces that cause the failure to occur. Figure 18.1 illustrates the effect of slope reduction in increasing the stability of a potential or existing landslide.

18.3.3 Structural solutions. Since the slope is unstable because of an excessive amount of force driving it to become flattened compared to the strength available to resist this force, a structural solution such as a retaining wall or other slope-reinforcing procedure may be utilized. The structural solutions add the strength that is missing within the soil mass to provide an adequate safety factor. Many of the same solutions discussed in Chapter 17 are available for slope stabilization. However, if a slope is actively failing, some engineering creativity is necessary in construction sequencing to enable the structural solution to be built without worsening the slide and destroying the proposed structure before it is completed. Sometimes this can be accomplished by alternating the construction areas (checker boarding) or other temporary measures.

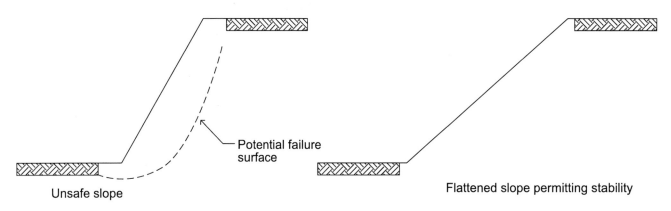

FIGURE 18.1
EFFECT OF SLOPE REDUCTION

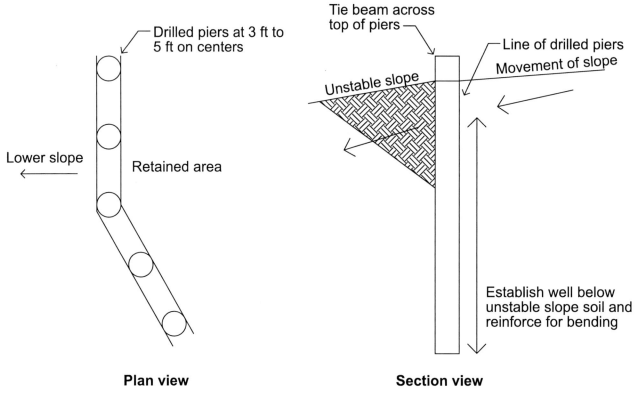

For SI: 1 foot = 0.305 m.

FIGURE 18.2
USE OF PIERS IN SLOPE STABILIZATION

Sometimes a slope can be stabilized by pinning it, either by drilling vertical or angled piers well into the ground past the failure plane of the slope or installing drilled-in and grouted tiebacks or other forms of soil reinforcement. See Figure 18.2.

18.3.4 Chemical stabilization. Unstable soil slopes can often be made stable through the injection of chemicals, such as lime or cement, that strengthen the potential slip surface within the soil. This is a highly specialized field that requires geotechnical studies by specialists who are experienced in the use of this approach for slope stabilization. For example, if effectively done, lime stabilization can transform CH clay into a CL or even ML soil by changing the mineralogical composition. If this can be effectively and uniformly done in the zone of high shearing stress, the soil will become stronger, and the slope may be stabilized.

18.4 TEST QUESTIONS

MULTIPLE CHOICE

1. A slope may become unstable because of (select two):
   a. seismic events
   b. seepage water
   c. tree planting
   d. traffic vibrations within 1000 feet

2. Slope reduction to stabilize an unstable slope involves (select one):
   a. removing soil from the bottom portions of the slope
   b. removing soil from the head of the slope
   c. removal of topsoil

3. What are the three broad categories of slope instability listed in this chapter? (select three):
   a. erosion
   b. downhill creep
   c. toppling
   d. seismic events
   e. landslide
   f. catastrophic failure
   g. devastating loss of a structure

4. What four procedures are discussed in this chapter to stabilize an unstable slope (select four):
   a. sub-drainage
   b. gunnite
   c. slope reduction
   d. rip-rap
   e. tree planting
   f. structural
   g. chemical
   h. piling
   i. paving

5. Stabilizing a slope by drilling vertical or angled piers well into the ground, past the failure plane, is often called (select one):
   a. hole stabilization
   b. pinning
   c. using stiff legs
   d. using dead men

6. Lime stabilization when used in stabilizing slopes can transform CH clay into a stronger _____ soil by changing the mineralogical composition (select one):
   a. MH
   b. OH
   c. CL
   d. gravelly
   e. cemented sand

7. When attempting to stabilize an active landslide, the engineer can use a technique that is called (select one):
   a. moving roll
   b. water jetting
   c. checker boarding
   d. forced drying

8. Chemical injection to stabilize a slope may use (select two):
   a. sodium chloride
   b. hydrocarbons
   c. cement
   d. surfactant
   e. lime
   f. sulfuric acid

Chapter 19

RECAP OF SITE AND FOUNDATION "RED LIGHTS"

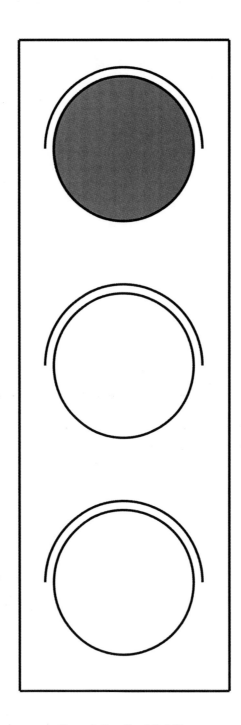

Foundation "red light"

Chapter 19

RECAP OF SITE AND FOUNDATION "RED LIGHTS"

19.1 Site Red Lights

Certain indicators in geotechnical reports or based on local knowledge are flashing red lights showing that there is a dangerous situation that should be addressed and the reports or design procedures should be carefully scrutinized. A "dangerous" situation could range from simply excessive settlement or heave causing an uncomfortable amount of cosmetic distress to a major failure of the structure, possibly endangering life and certainly endangering property. The following indicators are reasons for careful review and study of the site conditions.

19.1.1 Low bearing capacity. If the site contains soils that have low strength, the engineer must give careful consideration to the allowable bearing capacity used for various configurations of footings, whether they are shallow spot-and-strip footings, stiffened rafts and mats, or deep foundations. Low bearing capacity indications that may lead to shear failure of shallow or deep foundations include soils that have unconfined compressive strengths of 1500 psf or less, even in isolated areas, or a recommended safe bearing value of 1500 psf or less. These certainly deserve a second look.

19.1.2 Settlement. If the site has considerable depths of fine-grain soils that have water contents nearer to the Liquid Limit than the Plastic Limit, settlement should be a concern. Sites with organic deposits or landfills can cause a major settlement problem and should be carefully reviewed.

19.1.3 Existing uncontrolled fill. Uncontrolled fill can be fill that is clean soil and rocks but has been placed with unknown compaction and quality-control procedures. The fill in this case should be evaluated in a geotechnical report and a determination made if it is suitable to support a foundation, or if it can be made stable by re-work, or if the foundation should penetrate through the fill to something more stable. The worst case is an old landfill or "dump," which can settle several feet, sometimes years after placement, and in addition can produce toxic or flammable gas. These situations should be carefully studied.

19.1.4 Slope stability. Clay sites that have natural slopes steeper than 6 percent should be considered potentially unstable. Unstable could mean downhill creep or a landslide. The site foundation plans should provide remedies if such conditions are confirmed by geotechnical studies. The geotechnical report should contain discussions and, if appropriate, recommendations concerning potentially unstable slopes.

19.1.5 Dangerous seismic conditions. A review of Section 1613 of the 2012 IBC may indicate that a site could sustain devastating loss or collapse of structures or slopes under earthquake conditions. The code requires evaluation of the soil types related to different seismic regions.

19.1.6 Frost heave. If no consideration is present in the geotechnical report concerning frost penetration depths or measures to counteract the effects of frost heave in areas of significant frost penetration, the author should be questioned. Sometimes these reports are done by out-of-state firms or inexperienced engineers who simply forget that frost can be a problem in certain areas of the country.

19.1.7 Constructability – piers or piles. Recommendations for drilled piers or driven pilings should be evaluated with regard to constructability. One serious condition is large boulders in fill or naturally-occurring soils, which will make drilling piers or driving pilings without deflection or shattering very difficult. Another condition is a caving hole, which may influence the choice of drilled piers because of the requirement for casing and processing the hole to get the shaft through this material. If these situations are not anticipated, major schedule and cost overruns may occur, leading to change orders or attempts at contractor shortcuts.

19.1.8 Proposed significant cut or fill. If the site grading plan indicates major cuts or fills, the geotechnical report should have considered this factor in the recommendations made. Nearly any depth of cut and fill can be problematic, but certainly if the cuts and fills exceed 4 feet, careful consideration should be given. In expansive clay, cutting to any significant depth reduces the confining pressure and can generate excessive heave on the cut section. Likewise, fills, if established too steep, may not have long-term stability with regard to lateral movement, or if not properly compacted, may just simply settle under their own weight or the weight of the structure. Retaining structures may not be properly designed or constructed to deal with cuts or fills.

19.1.9 Groundwater problems. If it is known from local experience or from indications in the geotechnical report that groundwater can be a problem, it should be considered in the design and the constructability considerations for the project. A major problem can occur if unsuspected groundwater appears during construction causing schedules and budgets to escalate. In addition, below-grade areas, if subjected to groundwater rise, may cause the finished interior areas of the building to suffer water penetration with resulting mold and damp-rot problems as well as considerable irritation on the part of the occupants.

19.2 Foundation Red Lights

The following checklist indicates possible problems with the foundation plans:

- design not in accordance with geotechnical recommendations

- foundation location differs from site investigation location

- foundation plans not adequate or not complete

- low bearing capacity not recognized in footing design

- compressible or expansive soil on site not anticipated in design

- water seepage not anticipated in design

- seismic conditions not documented properly in design or calculations

- provisions for dealing with frost heave not included

- retaining structures or basement walls without adequate drainage

- under-floor crawl space not vented or drained.

19.3 TEST QUESTIONS

MULTIPLE CHOICE

1. Clay sites which have a natural slope of _____ or greater may be a problem (select one):
    a. 20 percent
    b. 6 percent
    c. 25 percent

2. Section 1613 of the 2012 IBC discusses sites that (select one):
    a. could have undesirable flooding
    b. could pose a seismic risk
    c. could be subject to damaging wind forces

3. Frost heave is a problem in localities such as (select one):
    a. Denver, Colorado
    b. Tampa, Florida
    c. Brownsville, Texas

4. Constructability of piers or piles (select one):
    a. should not be a concern when the engineer is reviewing design documents for a project since this is the contractor's problem
    b. should be considered in the design
    c. should not be a concern since constructability issues only add a small amount of time to the project

5. Groundwater will (select one):
    a. never be a problem on a construction site
    b. not concern the designers since it will be the contractor's problem
    c. always need to be considered in the design

6. Foundation plan red lights could be that (select two):
    a. the foundation location differs from site investigation location
    b. the foundation design is not responsive to the geotechnical report
    c. the plans are only one sheet long
    d. the applicable building code is stated on the plans

7. Sites with unconfined compressive strengths of _____ psf or less are a site red light (select one):
    a. 5000
    b. 3000
    c. 4000
    d. 1500

8. Sites with organic deposits or landfills can cause (select two):
    a. a settlement problem
    b. a problem with flammable or toxic gas
    c. unpleasant sounds
    d. displacement of homeless people

9. Methods of dealing with cuts or fills exceeding _____ feet should be reviewed (select one):
    a. 2
    b. 10
    c. 4

Chapter 20

CONSTRUCTION INSPECTION

Foundation inspection

Chapter 20

CONSTRUCTION INSPECTION

20.1 Elements of Construction Inspection

The best geotechnical investigation reports and the best foundation plans do not matter if the construction is not completed in accordance with those documents. Some contractors are not skilled enough to follow the plans, and a small minority deliberately do not follow the plans, attempting to reduce their construction costs and make a better profit from a fixed-priced contract. A good foundation or earthwork construction project consists of three principle elements: 1. an adequate geotechnical report; 2. proper plans that are responsive to the report and the site conditions; and 3. proper inspection of the construction. All inspections should include written reports distributed to all concerned parties. Governmental inspectors or private engineering companies may perform inspections on site preparation, excavation, fill operations, and concrete placement. Periodic or continuous inspections are required by the 2012 IBC through the Special Inspections process; see 20.7 for more detail.

20.2 Approved Drawings

Adequate inspection cannot begin until an approved set of construction drawings is available to the inspector. Often foundation plans reference the geotechnical report for certain aspects of site preparation; therefore, the inspector should also have this document available. The inspector also should be certain that the contractor's superintendent has the same documents available on the site and is following them.

20.3 Critical Points

20.3.1 Site preparation. Foundation plans often require certain site preparations or will refer to the geotechnical report for this feature. Site preparation always includes stripping of vegetation, topsoil, and trees. Tree roots should be grubbed. Site preparation could include the requisite slopes of fills and cut faces, placement of select fill, removal of unsatisfactory soil and replacement with select fill, and compaction. The site should be prepared so that surface water will not accumulate within the work and be properly drained off should a heavy rain occur. If subsurface drainage requirements are in the plans or in the geotechnical report, the drainage must be integrated, either before or during the construction of the foundation. The proper preparation of the site for foundation construction should be considered as important as supplying the proper strength of concrete or reinforcing steel, since often the site preparation is an integral structural part of a foundation and is critical to its proper performance.

20.3.2 Pier inspection. It is important that the inspector determine that the piers are to the required depths and are established in proper bearing material. Often this requires the geotechnical firm to send a representative to the site to verify the bearing material at the base of the pier. The piers may have to be deepened or possibly shallowed-up due to site conditions, and these situations should be confirmed with the geotechnical engineer. If groundwater or caving conditions are expected, the pier contractor should have available the right equipment and temporary casings as necessary to properly install the piers. Foundation pier contractors will attempt to avoid casing a pier hole if at all possible and may attempt to eliminate this very expensive process. A properly prepared set of construction documents will include an add-payment item for casing if needed, as well as add or deduct for deeper or shallower pier depths per lineal foot. It is important that drilled piers be placed in a dry condition and that caving hole conditions are properly controlled. In some cases placement under water or slurry by use of a tremie to the pier bottom for the concrete placement is permissible, but this should be done after consultation with the structural engineer.

Items to be recorded during pier inspection include the depth of pier establishment, the material at bearing depth, notations about confirmation of the material by a representative of the geotechnical engineer, and the proper placement of reinforcing steel including spacers or wheels to keep the steel within the typical 3-inch concrete cover range for separation from the soil. Concrete placement must be monitored during pier installation, and the structural engineer's specifications should advise whether or not the concrete can be permitted to free fall through the reinforcing steel or will have to be placed using a tremie pipe to get the concrete to the bottom as the hole is being filled. With deep piers, concrete falling through the reinforcing cage can be segregated, with the coarse aggregate being separated from the paste of the concrete mix.

If temporary pier casings are used because of caving soil or groundwater, the concrete should be placed at a slump of 6 to 8 inches to facilitate pulling the casing without lifting the steel cage. Increased slump requires more water in the mix, and more cement may be needed to maintain the required water / cement ratio.

If underreams (also known as "bells") are specified for the piers, the dimensions of the underreams should be verified by whatever means are necessary prior to the reinforcing cage being placed in the pier hole. One way to verify bell diameters is to measure the vertical movement of the "kelly" bar to produce the desired amount of extension of the wings of the belling tool. This movement can be repeated in the belled excavation to verify the diameter. Loose pieces of soil or soft soil should not exist at the bearing level in the underream or at the bottom of a straight pier shaft, and the pier-drilling contractor may be required to clean out the hole several times to achieve this. A strong light or mirror (if there is sunlight) can be used to confirm a clean bottom.

If the material in which the pier is stopped is unstable, it may be impossible for an underream to be cut for the pier, and the structural engineer should be consulted concerning deepening the pier or using some other methodology.

20.3.3 Piling inspection.
Small, routine driven-piling jobs may be checked based on local experience as to depth of penetration and final driving energy applied to the top of the pile. Care should be taken to ensure that the tops of wood piles are not "broomed" out due to the action of the hammer, and a driving cap should be used. In certain soil formations, including those with boulders, the piles may be deflected or even shattered, thus becoming useless, and careful attention should be paid to this possibility.

For major work with deep piling, it is best that one or more test piles be installed, with the driving energy at the advancing depths of the pile recorded. The pile should be tested by loading to produce a load-deflection curve that can be used to verify the design assumptions and provide guidance for the job service piles. See 13.3.2. Information recorded during service pile driving includes the type of piling being in accordance with the plans, the driving energy, and the depths of establishment.

20.3.4 Spread-or strip-footing bearing

Shallow footings, such as spread or strip footings, should be inspected to ensure that the soil bearing surface has the strength assumed in the design. The geotechnical engineer or a testing lab may have a representative present to verify that the proper bearing material is present and has adequate strength. If the design calls for a granular fill to enhance the bearing capacity below the footing, the placement of this material should be observed to ensure that the proper type of material with proper compaction is being used prior to forming up the footing on the surface of such material. If a footing is over-excavated, one of two approaches can be utilized: either the entire footing can be deepened to the excavated level and the bearing at that location verified; or the subgrade can be bought up with CLSM, lean concrete, or as otherwise specified on the structural plans. The bearing surface of the soil should be level or stepped in accordance with 11.4 and be clean of loose material and water prior to the placement of the concrete.

20.3.5 Structural configuration.
Proper dimensions of the footing should be verified against the plans, and proper placement and size of the reinforcing steel should also be verified. The location of bearing walls or plinths on the footing should be checked to ensure that unexpected eccentricity is not introduced into the foundation element.

20.3.6 Reinforcement inspection.

20.3.6.1 Rebar.
Reinforcing steel is used in all concrete footings but the smallest spread footings or certain types of footing walls and should be placed in accordance with the

designer's plans. If plans call for rebar to be near the bottom of the footing, that is where it should be, and not in the middle or the top. The bars should be cleaned of grease, rust, or dirt for proper bonding with the concrete. Steel bars should be tied in place to maintain their position, and chairs should be used to securely locate the bars' vertical position. The rebar should be secured in such a way that the concrete placement will not disrupt their location. Proper splice laps of rebar should be in accordance with the ACI Code and the structural plans. If there is a question concerning this, the designer should be consulted. It is important that the constructor maintains the correct concrete cover over the steel to prevent corrosion. This is typically at least 3 inches for all concrete in contact with the earth. Figure 20.1 is a photograph of a poorly set-up rebar foundation.

FIGURE 20.1
A POORLY SET-UP REBAR FOUNDATION NOT READY FOR CONCRETE PLACEMENT

20.3.6.2 Post-tensioning reinforcement.
Many foundations for small to mid-size buildings, especially for residences, are constructed with post-tensioning reinforcement. The reinforcement should be placed according to the plans and supported on chairs in order to be in the proper position within the slab and in grade beams or stiffener beams. Post-tensioning reinforcement is not intended to bond to the concrete throughout its length and should be covered with an adequate sheath of plastic, which will prevent it from bonding. The effectiveness of the reinforcement depends on the dead-end and live-end anchors, which are at the outer edges of the concrete. These anchors should be installed in such a way that they are positioned properly

and attached to the forms. The live-end anchors, at which the stressing force will be applied, will typically have a pocket former, which should be securely and tightly attached to the forms to exclude concrete from getting into the pocket void and later interfering with the stressing of the cables. The specifications will typically show details on how the reinforcement is to be placed and may include specifications referring to documents published by The Post-Tensioning Institute. For example, one item that may be contained in these documents is the necessity for taping any exposed cables so that bonding with the concrete does not occur.

Concrete used in all concrete footings or walls should be consolidated using vibration to place the concrete properly around reinforcing steel. This is especially important at post-tensioning anchors so that a "blowout" does not occur during stressing. A typical $^1/_2$-inch, high-strength, stranded-steel, post-tensioning tendon is stressed to about 33,000 pounds after the concrete hardens sufficiently, and any voids or weak concrete will become manifest immediately, sometimes explosively. After losses due to tendon friction and seating movement to engage the live end anchor jaws, 26, 600 pounds remains in each tendon for the life of the foundation to provide the pre-stress force. The tendons should not be cut without proper precautions to avoid injury or damage. This could happen if plumbing needs to be relocated and plumbers may not recognize the problems that could be created. If a tendon is cut, it should be repaired by a specialist. Figure 20.2 shows "honeycombed" concrete.

FIGURE 20.2
BADLY "HONEYCOMBED" CONCRETE APPARENT
AFTER FORM REMOVAL DUE TO INADEQUATE
VIBRATION DURING PLACEMENT

After the concrete has hardened sufficiently and the forms are stripped, the stressing of the post-tensioned cables (tendons) can commence. The concrete should have achieved a minimum strength, usually specified on the plans, prior to this action. The post-tensioning cables are pulled by a hydraulic ram, and it is important that the

inspector records both the elongation of the cable (stretch) as well as the hydraulic pressure utilized to stress the cables. The cable-stressing contractor should have a calibration chart available that indicates the force generated by a particular ram configuration at various hydraulic pressures. This calibration should be no more than six months old and should apply to the specific equipment (by serial number) being used. It is best to have the stressing verified by an independent inspector. The stressing contractor may be tempted to modify the stressing report if some tendons are not correctly stressed. It is very difficult to check this after the fact.

Once the cables are stressed properly with adequate cable elongation (related to the length of the cable), and the seating jaws have secured the cables, the cable ends can be cut off and the remaining void grouted to protect the anchorage and cable from corrosion. Locations near ocean coasts may have salts that will accelerate corrosion, and special caps may be needed. A proper inspection report will include an entry for each cable stating the stressing load prior to seating and the elongation of each cable after seating compared to the required elongation. Post-tensioned cable runs should not be less than 12 feet long because the stress loss from the seating relaxation becomes too large a percentage of the seating stress. Figures 20.3 through 20.8 illustrate the various steps in constructing a post-tensioned foundation.

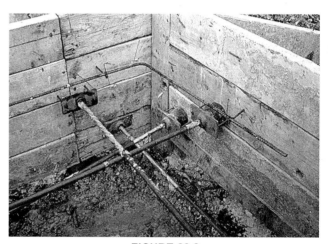

FIGURE 20.3
POST-TENSIONING CABLES SET UP IN FORM. LIVE OR
STRESSING END ANCHORS ARE ON THE RIGHT. DEAD
END ANCHORS ARE ON THE LEFT. MISSING CABLE
SHEATH IS REPLACED WITH TAPE.

Improper construction can ruin a well-designed foundation. Figure 20.6 shows a moisture-retarder membrane wrapped around pre-stress cables and cutting through the stiffener beam concrete. Figure 20.7 shows poor practice in placing concrete in a partial layer in a slab; a major horizontal cold joint resulted. This is not a monolithic concrete placement and is not acceptable.

FIGURE 20.4
STRESSING POST-TENSIONING CABLES AFTER FORM REMOVAL, USING HYDRAULIC RAM. INSPECTOR IS RECORDING MAXIMUM RAM HYDRAULIC PRESSURE.

FIGURE 20.6
UNCONTROLLED MOISTURE MEMBRANE WRAPPED AROUND PRE-STRESS CABLES AND CUTTING THROUGH STIFFENER BEAM CONCRETE

FIGURE 20.5
PAINT MARK ON CABLE AND ORIGINAL MARKING BLOCK TO PERMIT MEASURING CABLE ELONGATION AFTER STRESSING

FIGURE 20.7
CONCRETE WAS PLACED IN A 2-INCH LAYER AND THE PLANT BROKE DOWN. THE CONCRETE IS DRYING, AND A MAJOR HORIZONTAL COLD JOINT IS BEING FORMED.

20.3.7 Concrete quality considerations. Concrete mix designs with backup data should be submitted and approved prior to use on the job. Submittal review and approval is typically done by the structural engineer, often with input from a testing laboratory. The mix design will typically show pounds per cubic yard of coarse aggregate, fine aggregate, Portland cement, water, volume of chemical admixtures, and percentage of entrained air. The submitted mix designs should show the constituent properties of the mix and include proof of the strength of the mix to meet the specifications. The inspector should be supplied with batch tickets for each concrete truck showing the weights of the concrete constituents and adherence to the approved mix design. These tickets should be obtained as each truck arrives.

Proper control of the concrete should be provided to be sure the proper mix design is used, the specified slumps are maintained, and the concrete does not stay in the ready-mix trucks for too long.

A testing lab should be available to obtain test cylinders of the concrete and to verify the slump and air content. These procedures are typical for all job concrete, including pavement. Sometimes beams are cast for flexural testing on concrete paving jobs. See Figures 20.8, 20.9, and 20.10.

The cylinders are picked up by laboratory personnel after 24 hours, cured in the laboratory under standard conditions, and tested for compressive strength, typically at 7 days and 28 days of age. Structural plans should state a specified strength, such as 3000 psi compressive strength at 28 days. Test cylinders should be stored on the site for the first 24 hours in a level state, not subjected to disturbance, and protected from freezing.

FIGURE 20.8
INSPECTOR OBSERVING THE TYPE OF CONCRETE IN READY-MIX TRUCK. BATCH TICKETS SHOULD ALSO BE OBTAINED TO VERIFY THE CORRECT MIX IS DELIVERED AND TO COMPLETE THE RECORD.

FIGURE 20.9
SLUMP TEST ON FRESH CONCRETE. SLUMP IS SPECIFIED ON PLANS OR IN THE JOB SPECIFICATIONS.

Typically concrete should be placed within 90 minutes of having water added at the concrete plant. Excessive water should not be added on the job site, and if this is required continuously in quantities greater than about five gallons per truck load, the mix design and plant batching procedures should be investigated. If air temperatures are 90°F or higher and the humidity is less than 25 percent, the concrete may suffer drying shrinkage cracking. Concrete should not be placed if air temperatures are 40°F and falling unless special cold-weather measures are in place. Refer to the *Concrete Manual* published by the International Code Council for detailed information about concrete properties, mix designs, and hot or cold weather procedures.

Curing of the finished concrete is recommended to improve strength and reduce surface dusting and drying-shrinkage cracking. Curing may be by fog-water sprayers, moistened mats, or sprayed on curing compound.

20.3.8 Finished grades and drainage. In all phases during foundation construction, surface water should be directed away from the foundation area and not be permitted to stand within or adjacent to the work. The same is true after the work has been completed. The desired finished grades will typically be noted either on the foundation plans or the site plans. Foundations must have an adequate concrete reveal of 4 to 8 inches before the top of the soil is encountered, and the areas around the foundation must slope away to a swale or drainage pipe system that will drain adequately off the site. For certain types of construction, such as shopping centers and other commercial buildings, these elevations and grades should be shown numerically on the site development plan. Slopes should have an approximately 6-inch fall within 10 feet from the building on all sides for all buildings, including residences. Concrete flatwork or pavement may have lesser slopes, and certainly door thresholds will have to be designed specifically for access ease as well as to prevent water entry into the building.

FIGURE 20.10
STRENGTH TEST CONCRETE CYLINDERS BEING CAST ON JOB SITE

20.4 Earthwork Inspections

20.4.1 Site preparation. Before earthwork is undertaken, the site must be cleared of topsoil, grass, bushes, trees, old structures, or old pavement. This is typically done by utilizing blade equipment, such as a bulldozer or a grader. Trees should be removed and the stumps and major root systems also removed. Voids left by stump removal should be filled with acceptable soil using controlled compaction. All material cleared should be removed from the work area to avoid being mixed with the new controlled-density fill. The topsoil could be stockpiled for use later by landscapers if otherwise acceptable to the specifications. Burning of trees and bushes is sometimes utilized by contractors to dispose of this type of vegetation on site to avoid expenses of hauling it away and paying a fee to dump it in a landfill. However, some local regulations may prevent site burning, especially during periods of dry weather. It is never acceptable to bury vegetation within a future use fill since it is likely to decompose with time and create a settlement issue. Also, site preparation is a good time to determine if there is an unrecorded landfill or dump on the site, in which case operations would have to cease until this problem is cleared up properly.

20.4.2 Elevation control.

The finished elevation of earthwork must correspond to that desired by the site civil engineer and as shown on the site grading and drainage plan. The engineer establishing elevations of earthwork should consider that topsoil or pavement sections may be placed above the finished compaction controlled soil. The site grading and drainage plan will typically show existing contours before the work is started, usually in dashed lines, and will show the proposed final elevations, usually in solid contour lines. In addition, spot elevations may be shown relative to mean sea level, project datum, or site benchmark referenced on the plans. The contour lines on these plans indicate locations of the same elevation wherever the line leads. For example, an elevation of 950.0 would be represented by any line that could be followed across the site at elevation 950.0. As the contours become closer together, the slope is indicated as becoming steeper, and vice versa. Figure 20.11 illustrates a segment of a site grading and drainage plan with existing and finished contours plus spot elevations in various locations. Figure 9.1 is a subdivision grading and drainage plan.

Elevation control is accomplished by the use of cut-fill staking. This method refers to placement of an elevation point on the surface of the ground with a lath next to it, which indicates how much cutting or excavation, or how much filling needs to be done at that point to obtain the desired final grades, allowing for material to be placed over the graded soil surface, such as topsoil or pavement sections. Refer to 9.12 for a more detailed explanation of cutting or filling controlled by cut-fill stakes. Cut-fill stakes are normally set by a surveyor or a skilled construction person.

20.4.3 Slope measurement.

The earthwork inspector may need to determine the slopes of fills or cuts on the site to meet the plan or code criteria. The slope can be measured most accurately by the use of a survey crew who will do a standard elevation and distance survey and will compute the slope. Slopes can be expressed as a percentage or as a ratio, such as 5 percent or 1v:20h. Approximate slope determinations in the field can be done in several ways using a hand-held slope inclinometer or a hand-held level. These are inexpensive pieces of equipment and consist of an eyepiece and a leveling bubble. In the case of the hand-held slope inclinometer, the eyepiece can be aimed up or down the slope, then compared to a horizontal plane as shown by the reference bubble in the device. The bubble, if centered, will indicate that the instrument is level, much like the bubble in a carpenter's level. The slope inclinometer will have a unit that can be held in a horizontal position with a bubble, but will have a movable part that can be rotated until it is parallel to the slope; the slope read off of a vertical circle on the equipment.

The inspector employs a hand level by determining the distance from the ground to her or his eye, for example 5.5 feet, and, standing at the base of the slope or somewhere in the middle of the slope, as convenient, uses the hand level to observe a point level with his or her eye up the slope. The inspector then marks that point and measures the horizontal distance to where he or she was originally standing. The distance must be measured horizontally and not following along the slope of the ground, which would distort the distance somewhat. The inspector can do this by either securing the end of a tape on the ground at the point which he or she spotted or by having an assistant hold the tape where directed, and then extending the tape out to the position of the observation, making sure the tape is kept level and not sloping. The slope can be determined by dividing the vertical distance, in this case 5.5 feet, by the horizontal distance, which could be 100 feet. A simple calculation will show that this is a 5.5 percent slope. By dividing the horizontal distance by the vertical distance, the inspector can obtain a ratio of slope, which is a common way of expressing it; in this case it is 1v:18h.

The slope inclinometer is similarly used in that the inspector selects a post or fence or point on a tree up the slope or down the slope at the same height above the ground as the inspector's eye, for example 5.5 feet. Using the slope inclinometer to sight on this point and keeping the reference portion level based on the bubble, the inspector can directly read the slope off of the vertical circle. If the criterium for the slope is 2 percent, the desired slope would be either 2 percent or 1v:50h.

If the desired slope is 5 percent, it could be noted as 1v:20h. The measured slope may be compared to the desired slope; for example, if the slope is 1v:18h (measured), the slope is too steep. If the readings of the slope and calculations indicated 1v:25h and the desired slope is 1v:20h, the slope is too flat.

Similar procedures could be applied to the grading and drainage plan if it is desired to determine the plan slope at a particular location. Select a contour with a given elevation, say it is 550, and another contour further down or up the slope with a given elevation, such as 560, and determine the horizontal distance between by means of scaling the plan using an engineer's scale and the scale that is on the drawing. The slope can be calculated using the same calculations as described above.

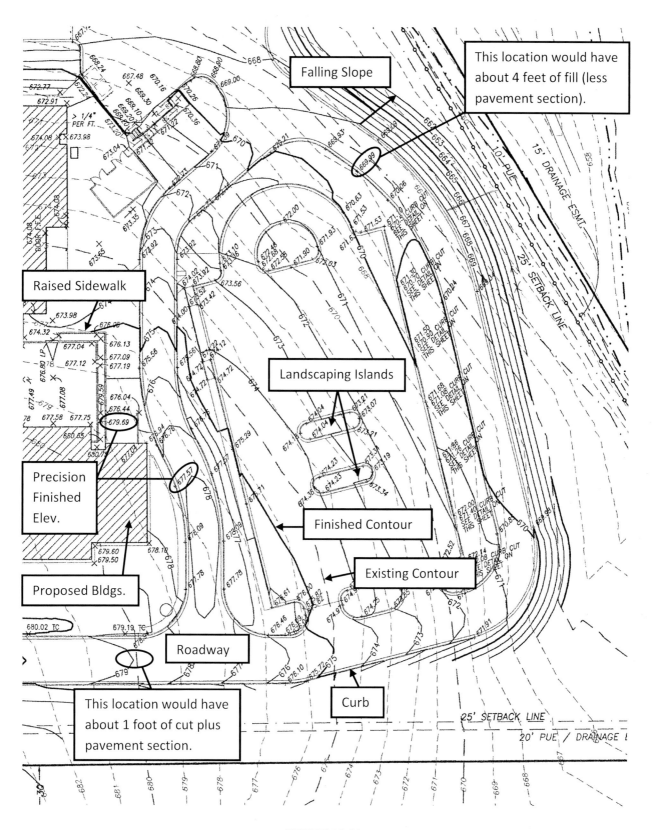

FIGURE 20.11
PORTION OF SITE GRADING PLAN

20.4.4 Soil material control. Specifications may include criteria for the type of soil to be used in certain places, such as select fill or flexible pavement materials. Select fill is often specified beneath building foundations, and the drawings or the geotechnical report should indicate the specification criteria for this material. This criteria almost always includes the Atterberg Limits and the grain size distribution (Sieve Analysis). To properly determine the material quality, a testing lab should be asked to take one or more bulk samples of the material from a job-site stock pile, offsite stock pile, or from the job fill placement itself and run laboratory tests, which include as a minimum the Atterberg Limits and the grain-size distribution. Sampling for fill material can be from a pit or from a stockpile. Active dirt pits usually have a vertical cut face. The supplied material often is based on a front-end loader cutting full height of the cut. Upper topsoil should be excluded from the face cut. Sampling should include a front end loader producing a local stockpile, which is then sampled as described below. Large stockpiles of select material found either at a pit or crusher plant are sampled by a front-end loader taking a full-height bite from four sides and creating a smaller representative pile. This pile is sampled from four sides with shovel samples deposited in a smaller pile, repeating until a manageable size sample can be obtained. Clean aggregate smaller stockpiles are quartered, combining opposite quarters until a manageable size sample is obtained. Cone-shaped stockpiles of aggregate formed by conveyor-belt discharge, power shovels, or draglines may have coarse aggregate segregated into an outside lower level zone. Therefore, deep bites with a loader are necessary to obtain representative samples. The final sample must be representative of the sample source. For full laboratory testing including Atterberg Limits, grain size, and Proctor Testing, about 100 pounds of sample is needed. The sample must be transported to the laboratory in buckets with tight lids or in bags with plastic liners to avoid loss of fines during transport. All samples must be tagged with the name of project, proposed use, source, date, and names of sampling personnel. The result of the testing can determine whether the specifications are met. If the specifications are met, remaining portions of the same bulk samples can be used to obtain the Proctor curves for density control as set forth in the following section.

20.4.5 Compaction control. If a required density is specified for general earthwork, select fill, flexible base, or other material from the job, and if the materials present meet the criteria as discussed in 20.4.4, field density testing is normally used to verify compaction. Field density tests are almost always done with a nuclear meter, although the U.S. Army Corps of Engineers sometimes uses the sand-cone method, which, if properly carried out, is more accurate than the nuclear meter, but takes much more time to perform and is more expensive. Nuclear meters are generally accurate enough if properly calibrated to adequately control compaction of soil materials and HMAC. To control compaction one must know the specifications on the plans, in the project manual, or in the geotechnical report for the density required for various types of materials depending on the use on the site. These specifications usually refer to compaction standards, such as "Standard Proctor" or "Modified Proctor." The standard Proctor compaction criteria is used for most building earthwork and light pavement, while Modified Proctor is used for airfields or heavy construction, such as major highways. The 2012 IBC J107.5 calls for 90 percent of Modified Proctor for fills. The plan specifications or geotechnical report will specify a percentage of the maximum laboratory dry density to be obtained on the job. The nuclear meter determines the unit weight of the soil and compares it to the laboratory Proctor maximum density. The percentage of the maximum laboratory density is specified typically as 90, 95, or 100 percent for fill. The density testing will determine if the criteria is reached or exceeded in the field. If it is not met, additional work will need to be done to increase the compaction. Generally, this includes more rolling, and maybe adding moisture or drying out moisture and rolling the lift and testing it again. Refer to 9.5.3 for details of the Proctor test and other information about compaction of soils.

For all sorts of compaction control it is very important that the material being tested in the field is the same material for which a Proctor curve has been run and used for density comparison. Many construction errors have resulted because this condition was not satisfied. Many soils have a similar appearance of color and texture, but may be somewhat different in their plasticity or grain size distribution and will have a different Proctor curve. If there is any question as to which Proctor curve is correct, the soil should be sampled and Atterberg Limits and grain-size distribution should be tested. These values should be reported on each Proctor curve done by the laboratory. A comparison should be made to see if the basic index properties and appearances are approximately the same.

Refer to 9.12 for photographs of typical earthwork equipment including compaction equipment. It is not the inspector's job to dictate to the contractor what type of equipment should be used to obtain the specification densities, but some knowledge of the function of the various equipment will be useful to the inspector.

20.4.6 Identification of cuts or fills on a site. There are two methods of identifying existing cuts or fills on a site: study of the site grading plan or field observations. Both methods are similar in that they require a comparison of original ground surface to finished ground surface. The original ground surface is indicated on the site grading plan (usually by dashed contours), and the finished grades are shown by solid line contours or spot elevations. If finished elevations are lower than the original elevations (contours), a cut is required. If finished elevations are higher, filling is required. For clarity, a cross-section could be plotted of the area of interest and the amount of cut or fill would be readily seen.

The same process is followed for field observations without the convenience of a site grading plan. The observer would need to visualize the original ground surface and note the existing final elevations. The original ground surface can be estimated by noting trees around a suspected fill that do not have their trunks partially buried or observing the depth of constructed tree wells. Also the original ground surface is likely to exist at a flatter slope than the slope of edges of a fill or an up-hill cut. The changes of elevation can be surveyed by using the techniques described in 20.4.3.

To confirm the presence of fill, test borings (see Chapter 6) or excavated test pits can be used, either by hand excavation or by backhoe equipment excavation. For shallow exploration, test pits are a better source of information. Test pits should be sampled at 1-foot intervals of depth. Usually disturbed samples are obtained. Careful inspection and logging of the sides of the pits along with detailed photographs are needed. Location and elevation of the pits should be recorded. Undisturbed samples of fill or original soil are difficult to obtain from test pits, but could include 6-inch blocks carefully cut from pit walls or a variation of the Shelby tube sampling procedures described in Chapter 6. The best approach is for the engineer to carefully inspect and log the walls of the test pit after cutting away the smeared soil from excavation processes. Fill can be identified by the presence of man-made materials (plastic, lumber, concrete pieces) or by finding the old topsoil line at the bottom of the fill. Topsoil is almost always a dark soil layer, possibly with grass and roots. If the fill was placed using engineered fill procedures, the topsoil line may not be present. Other indicators of fill are the lack of soil layers, or structure that can be found in undisturbed soil, or finding fill composed of material that is different from the undisturbed soil.

Non-engineered fill may be obvious, containing boulders, brush, or trash. However, on sandy or silty sites, the fill may look a lot like the native soils, and if the indicators described above are not seen, identification of fill by observation may be problematic. If there is confusion about the presence or quality of fill, geotechnical professionals may need to be employed.

Cut-fill sites are usually constructed by cutting on the uphill side and using the excavated material to fill the downhill side. Fills that were not placed in a controlled manner may settle, causing structure damage. If the soil is an expansive clay, swelling due to overburden unloading of the cut side as well as settlement of the fill side may both occur. Figure 20.12 illustrates cross-sections of a fill only site and a cut-fill site. A cut-fill site may also be called a transition site.

Figure 20.13 shows fill on a site found by observations. Boulders are visible and the fill is likely uncontrolled.

20.5 Inspections of Small Roadways and Parking Lots

20.5.1 Small roadways.

20.5.1.1 Geometry. Roadway geometry is normally specified by an engineer's plans or controlled by state, county, or municipal regulations. Usually the roads under the control of these governmental entities are classified depending on use and traffic, with the smallest roads being used to service single-family residential developments. Minimum paved widths for suburban and urban low volume roadways will typically be 30 feet of pavement, which will provide for on-street parking as well as a traffic lane. This usually requires a right-of-way of 50 feet to provide a 10-foot area on each side of the pavement in which utilities are often placed. In urban or suburban areas, as well as county roads, 60 feet or more of right-of-way is required to accommodate the drainage ditches on each side of the roadway if a curb and gutter section is not used. For a low-speed pavement with only occasional truck traffic, 11-foot widths for each lane are typically used. If there are higher speeds proposed on the roadway and a significant amount of truck traffic, a minimum of 12-foot lanes should be required.

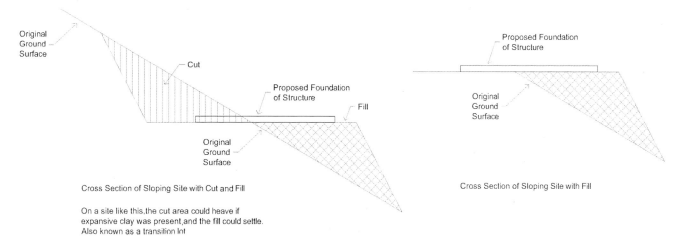

Cross Section of Sloping Site with Cut and Fill

On a site like this, the cut area could heave if expansive clay was present, and the fill could settle. Also known as a transition lot

Cross Section of Sloping Site with Fill

FIGURE 20.12
CROSS-SECTIONS OF A CUT-FILL SITE AND A FILL-ONLY SITE

FIGURE 20.13
FILL FOUND ON SITE BY OBSERVATION

Private drives serving remote single-family residences or apartments and condominiums may have smaller roadway widths; however, a minimum lane width of 10 feet should be utilized. In multi-family developments, some truck traffic must be anticipated for delivery and trash pickup. A minimum 12-foot lane should be provided for these functions. The minimum turning radius should be 50 feet measured to the edge of the pavement to accommodate truck turning. A cul-de-sac, which is a circular turn around at the end of a street, should have a minimum paved radius of 50 feet. These requirements are to accommodate fire equipment and delivery trucks. Even for private roads, the fire department will want to insure access for its equipment and will require the minimum geometry. In addition, the fire department will require a minimum structural strength of the pavement to provide all-weather access in case of fire or other emergency. Single-lane private roads should be a minimum of 14 feet in width plus shoulders to allow for vehicle passage around a stalled vehicle. If curbs and gutters are not utilized, 4- to 8-foot shoulders should be provided, usually by extending the base material out beyond the paved surface.

Typical roadways have a crowned cross section with the center being higher and sloping off to each side to permit discharge of water during rains. If traffic speed exceeds 30 MPH, the curves in the roadways should have the traffic lanes widened. Roadways with traffic speeds in excess of 40 MPH often have curves super elevated, meaning the outer edge of the pavement is raised higher than the inner edge of the pavement to permit reduction of the centrifugal force on vehicles.

20.5.1.2 Drainage. Drainage is very important for roadway construction, both from a point of view of safety and convenience during rain storms and to preserve the pavement structure itself. Drainage in curb and gutter sections is typically by curb inlets and a storm drain pipe system. Road sections without curbs and gutters will slope down from the edge of the shoulder at no steeper than 1v:4h into a drain ditch on each side of the roadway. These are some-

times called "bar" ditches. The ditches, as well as the pavement surfaces, should have the grades considered carefully to insure drainage occurs during a design rain storm. The bar ditches must be provided with a method of discharging into the countryside or a creek, and excess velocity should be considered with regard to erosion of the ditches themselves. If velocity is excessive, sometimes the ditch will be armored with concrete or stone.

Low-water crossings are sometimes designed for very low-volume roads, especially private drives. These types of water crossings should only be used as a last resort and clearly marked with signs showing the depth of water. A number of lives are lost each year in the United States due to vehicles being swept away at low-water crossings and the occupants drowned. If low-water crossings are utilized, they should be constructed of concrete to avoid being damaged by erosion from the high water.

20.5.1.3 Grading. Construction of a small roadway is quite similar to any other type of earthwork in that a finished grade elevation must be obtained by earthmoving equipment. The use of cut-fill stakes is generally required, although for very small projects the earthwork equipment operator's judgment may suffice. The earthwork should be brought to the subgrade elevations, which are the elevations below the pavement structural material. The subgrade should be scarified, compacted, and controlled by Proctors and field densities as described in Sections 20.4.5 and 9.5.3. The cut-fill stakes should indicate what is to be cut or filled to develop the center line, edge of pavements, edge of shoulders, the center line of bar ditches, and edge of right of way. An engineer's plans should be utilized to determine the finished elevations at these points and the cut-fill stakes and blue tops established accordingly.

20.5.1.4 Pavement structural sections. There are basically two types of pavement structural sections utilized: concrete or flexible pavement. Concrete pavement for low-use roads and driveways should not be less than 5 inches thick and supplied with a minimum amount of steel reinforcement to control temperature and shrinkage cracking. Concrete roadways should have a transverse expansion joint about every 100 feet. This is a joint that can move back and forth with changes in temperature. There is normally a dowelled connection in this joint, and one side of the dowel is embedded in the concrete. On the other side of the joint the dowel is free to move, usually by use of greased paper or other methods. The dowels are typically $5/_8$-inch smooth steel bars 18 to 24 inches long and located about 18 inches apart. These details should be illustrated in the designer's plans. In addition, contraction or control joints should be formed or sawed into the concrete transverse to the long direction of the roadway at about every 20 feet. Concrete roadways usually have a center line longitudinal joint, which is tied together by deformed rebar, but not in the form of an expansion joint. The type of the subgrade soil, a sub-base, and the concrete pavement thickness itself is subject to design by the roadway designer or by municipal regulations. The factors that are considered in the design of the pavement sections are anticipated volume and character of traffic and the subgrade soil conditions and types. Loadings from passenger automobiles are generally not the controlling factor for pavement structural designs, but design is based on wheel loads of trucks. Naturally, the greater the volume of truck traffic, the more intense the loading conditions are on the pavement.

A flexible pavement typically consists of a compacted subgrade soil and a base, either composed of crushed stone or lime-stabilized soil. The surfacing is usually 2 inches of hot-mix asphaltic concrete (HMAC). The base, depending on traffic volume and loadings and the soil conditions, is typically 6 to 12 inches thick. There are numerous variations in the pavement structural sections that may be utilized, usually depending on local economics and availability of materials. On low-use roads occasionally the surface is a seal coat, sometimes called a chip seal. This is a liquid asphalt sprayed on the surface of the approved base that has crushed rock rolled into the asphalt. The thickness of this surface is $1/4$ to $1/2$ inches.

20.5.1.5 Inspections.

Roadway inspections consist of the inspector verifying the geometry as described above either based on engineered plans or local regulations. This can often be done by the use of simple taping procedures to check widths or may require the use of a surveyor if more detailed criteria are to be met, such as curvature degrees and super elevation.

• The drainage features should be checked by some type of surveying procedure, which could be as simple as using a hand level or as involved as using a surveyor with surveying equipment.

• Subgrade should be verified to be to the cross sections required and compacted and tested by field density testing in conjunction with laboratory Proctors. Typically 95 percent of Modified Proctor is the density criteria, but the job specifications should be consulted. The base course may consist of graded crushed limestone or other rock, and this material should be tested by the laboratory to see that it meets the gradation and plasticity requirements of the specifications. The density of the compacted base should be tested by field density tests compared to a Proctor curve.

• Concrete pavement should be constructed after the approval of a submitted mix design from the concrete plant showing that the mix meets the specifications of either the engineer's plans or the local regulations. Typically 3,000 psi to 4,000 psi concrete is specified for concrete pavements.

• During placing of concrete, the thickness of the concrete should be determined based on the height of side forms or by a graduated rod inserted into the plastic concrete, and the quality of the concrete during placement should be verified. Refer to 20.3.7.

• If the base or sub-base for flexible pavement or concrete pavement consists of lime-stabilized soil, the per-

cent lime mixed with the soil should be determined during construction. Usually it is specified as a percentage based on dry weight, and some mathematical calculations can reveal the number of pounds of lime per square yard for a given thickness of stabilization that will be required. This can be verified by delivery tickets showing pounds of lime applied per square yard. Sometimes a liquid slurry mixture of lime is used, requiring a similar calculation. There is a difference in the potency of hydrated lime and quick lime, and the calculation must be adjusted accordingly. Quick lime can cause burns and must be carefully used. The procedures are somewhat complicated, and it will be necessary for the engineer or inspector to employ a testing lab to assist in this determination.

• The surface of flexible pavement is typically HMAC, which is applied and rolled hot on top of the approved base material after a liquid asphalt prime coat is applied. The specifications may vary somewhat, but usually the HMAC should be a minimum of 185° F at the time of placement and rolling. This should be verified by using an appropriate thermometer. The compacted thickness of the HMAC should be verified by measurement, and the density after rolling should be verified by nuclear meter readings, usually in conjunction with more widely spaced cores of the HMAC. These procedures will require the use of a testing lab.

• There are two commonly used procedures for design of HMAC in the United States, the Marshall Method and the Hveem Method. Both methods of design utilize the properties of the aggregates and the asphalt oils to produce a mix design, also called a job-mix formula. The laboratory must determine that the mixture will attain certain stability criteria and provide a reference density for the field density testing. The two methods have two different basis for reference densities, and a testing lab should be aware of which method is being used in the design. The required field densities are shown in the specifications or in the municipal regulations. Typical spacing of density tests, both for the subgrade, base course and HMAC would be a nuclear density test every 100 to 200 feet of lane length. For more important work, cores of the HMAC should be taken, typically one every 500 feet. HMAC coring may need to be preceded by an ice pack if the pavement is still hot after it has just been laid to prevent raveling of the cores.

• No water or ice should exist on the subgrade or on the base, and the HMAC should not be placed and rolled during rain or cold conditions. Minimum ambient and ground temperatures are found in the specifications and in the Asphalt Institute reference listed below. HMAC is usually placed by a lay down machine, which places the hot mixture at the required loose depth and does initial compaction on it. Additional compacting is done by steel-wheel and rubber-tired rollers. All rolling compaction should be completed before the HMAC cools too much.

- See *Construction of Hot Mix Asphalt Pavements*, Manual Series No. 22, available from the Asphalt Institute, Lexington, KY, for detailed information on mix design and placement.

20.5.2 Parking lots.

20.5.2.1 Grading. The earthwork involved in grading for a parking lot is similar to the process described previously in Section 20.5.1.3. Before the contractor places fill, the surface soil should be stripped of topsoil and vegetation and tree roots removed and grubbed. Any depressions left from this removal should be filled with compacted soil. The entire subgrade level should be scarified, moisture conditioned, compacted, and tested for density compliance prior to placing additional fill or pavement sections. Refer to Chapter 9 regarding benching for subgrade slopes that are to receive fill. Cutting and filling is controlled by cut-fill stakes or laser level controls to achieve the subgrade elevation. Fills should be tested for density compliance with the specifications, typically using nuclear density meters and Proctor curves developed in the laboratory for each type of soil. The moisture at the time of compaction should be near the optimum. The final fine grading of the subgrade with blade equipment, such as motor graders, should be done when the soil is near the optimum moisture content to avoid tearing the surface with the blade. This is especially important on clay soils. Most parking lots are laid out by using engineered grading and drainage plans, and these elevations should be followed during construction. It is important that the American's with Disability Act maximum grades are followed in the vicinity of handicapped parking spaces, which typically should not exceed 2 percent surface slope. The remainder of the parking area may slope up to 5 or more percent; however, on shopping centers in which shopping carts could be used, consideration should be given to prevent runaway carts, which could careen downhill and strike pedestrians or parked vehicles.

20.5.2.2 Drainage. Parking lots can produce large quantities of surface runoff during rains, and this should be provided for in the drainage plan. Plan details of curb and area inlets should be followed carefully to insure the inlets actually collect the surface water where they are required to do so. The sizing of the drainage features would be dependent on an engineered plan. Parking lots typically are equipped with curb and gutter edges, which will define the extent of the paving and control the runoff. During construction, provisions should be made so that water does not accumulate in the earthwork or paving sections and cause softening of the soil. Should water accumulate while the contractor is in the midst of earthwork or base material construction, examination should be made and the area may need to be excavated and re-constructed with dry and stable material.

20.5.2.3 Pavement sections. Pavement sections for parking lots have similar considerations to those described in 20.5.1.4, and the pavement may be concrete or flexible pavement. On larger parking lots, drive lanes are typically marked and should be used by delivery trucks and trash pickup trucks. These areas usually will have heavier pavement sections to accommodate the greater wheel loads. Areas reserved strictly for automobile parking may be of lighter construction. The parking lots are typically paint striped to indicate the parking spaces and drive lanes. On larger projects, traffic control is necessary by use of signs or islands, which will direct traffic.

Inspections would be similar to those described for small roads in 20.5.1.5. A testing laboratory will be needed to test the subgrade, sub-base, concrete, and base courses and HMAC as the case may be.

20.6   Inspection Reporting

Depending on the extent of the project, an inspector may be present continuously or intermittently. In either case, each inspection day should have a daily report generated that would indicate the project description or designation, contractors working, equipment present, and weather conditions. In addition, any testing laboratory activity should be noted with number of density tests, cores, or other activities. Cubic yards of concrete, truckloads of base, or tons of HMAC placed for each day should be noted. A plan should be kept by the inspector showing where the various placements of concrete or other types of pavement were done on each date. Approved concrete mix designs and HMAC mix designs should be obtained by the inspector and included in that day's report. A file of all laboratory reports should be maintained and organized by report date.

A narrative should be written in the report stating what occurred that day, if anything unusual was noted, and if the plans seem to require corrections in either geometry or section or drainage arrangements. If questions arise concerning these items, the inspector should contact the design engineer or his or her supervisor to see that the questions are properly transmitted for solutions.

A written record of soils observed and tested, rejected or accepted should be made. Earthwork and paving are ongoing processes, and generally the acceptance or rejection of compaction in earthwork or paving material can be orally given to the contractor's superintendent so the work can continue if results are acceptable. A written report should document the oral comments in all cases. Typically the general contractor on the project has a subcontractor in charge of the earthwork. At the beginning of each project, it should be determined who can be officially advised orally of conditions passing or failing; the earthwork contractor superintendent, the general contractor superintendent, or both. If there are serious or unsafe conditions being produced by the contractor, not only should the contractor and subcontractor be advised orally, but the design professionals and other interested parties should be notified immediately by telephone with written reports to follow.

The inspector may not have the authority to stop work on the project, especially if the inspector is not representing the permit authority. However, in all cases of noncompliance, the contractor should be advised immediately and reports produced as quickly as possible so that noncompliant material or work is not covered up and therefore the problem becomes much more expensive to alleviate in the future.

Reports should be distributed to the building official, the design professionals for the project, the project owner, and the contractor.

20.7 Special Inspection Program (Section R104.4 of the 2012 IRC and Sections 104.4, 1704, and 1705 of the 2012 IBC)

20.7.1 Special Inspection Program. The International Code Council (ICC) has published guidelines for Special Inspections titled "Model Program for Special Inspection" in conjunction with the International Accreditation Service (IAS). The following is extracted from these guidelines, which are not code requirements, but are model procedures that could be adopted by local building officials:

A. Purpose of Special Inspection

Special inspection is the monitoring of the materials and workmanship that are critical to the integrity of the building structure. It is the review of the work of the contractors and their employees to ensure that the approved plans and specifications are being followed, and that relevant codes and ordinances are being observed. The special inspection process is in addition to those inspections conducted by the municipal building inspector and by the design professional in responsible charge as part of periodic structural observation. Special inspectors furnish continuous or periodic inspection as prescribed in 2012 IBC Tables 1705.2.2, 1705.3, 1075.6, 1705.7, and 1705.8 for that construction that requires their presence (see 2012 IBC Sections 110 and 1704).

Good communication between the special inspector and the designers, contractor, and building department is essential to project quality assurance. The following section is a combination of building code requirements and the results of years of lessons learned related to special inspection work that have resulted in the successful completion of buildings.

B. Duties and Responsibilities of the Special Inspector

The IBC requires that a "Statement of Special Inspections" be submitted with the application of the permit. The special inspector should know and understand the scope of the statement prior to beginning special inspections (see the "duties of the design professional in responsible charge").

Though not required by code, special inspectors and/or inspection agencies can document acceptance of their responsibilities and the scope of work for a project by signing an agreement that includes a detailed schedule of services.

Duties of special inspectors and/or inspection agencies include the following:

1. General requirements. Special inspectors shall review approved plans and specifications for special inspection requirements. Special inspectors will comply with the special inspection requirements of the enforcing jurisdiction found in the Statement of Special Inspections, including work and materials.

2. Signify presence at job site. Special inspectors shall notify contractor personnel of their presence and responsibilities at the job site. If required by the building official, they shall sign in on the appropriate form posted with the building permit.

3. Observe assigned work. Special inspectors shall inspect all work according to the Statement of Special Inspections for which they are responsible for compliance with the building-department-approved (stamped) plans and specifications, and the applicable provisions of the 2012 IBC Section 1704.

4. Report nonconforming items (discrepancies). Special inspectors shall bring all nonconforming items to the immediate attention of the contractor. If any such item is not resolved in a timely manner or is soon to be incorporated into the work, the design professional in responsible charge and the building official should be notified immediately and the item noted in the special inspector's written report (see 2012 IBC Section 1704.2.4). Some jurisdictions may require this report to be a separate, individual report from the progress reports. The building official may require this report to be posted in a conspicuous place on the job site. The special inspector should include in the report, as a minimum, the following information about each nonconforming item:

   • Description and exact location.

   • Reference to applicable detail of approved plans and specifications.

   • Name and title of each individual notified, and method of notification.

   • Resolution or corrective action taken.

5. Provide timely progress reports. The special inspector shall complete written inspection reports for each inspection visit and provide the reports on a timely basis as determined by the building official. The special inspector or inspection agency shall furnish these reports directly to the building official and to the design professional in responsible charge (see 2012 IBC Section 1704.2.4). These reports should be organized on a daily format and may be submitted weekly

at the option of the building official. In these reports, special inspectors should:

- Describe inspections and tests made with applicable locations and whether the work meets the requirements of the Statement of Special Inspections.

- Indicate nonconforming items (discrepancies) and how they were resolved.

- List unresolved items, parties notified, and time and method of notification.

- Itemize changes authorized by design professional in responsible charge if not included in nonconforming items.

6. Submit final report. Special inspectors or inspection agencies shall submit a final signed report to the building department stating that all items requiring special inspection and testing by the Statement of Special Inspections were fulfilled and report, to the best of their knowledge, that all items are in compliance with the approved plans and specifications (see 2012 IBC Section 1704.2.4). Jurisdictions may also require the design professional in responsible charge to sign the report before it is submitted to the building official. Items not in conformance, unresolved items, or any discrepancies in inspection coverage (i.e., missed inspections, periodic inspection when continuous inspection was required, etc.) should be specifically itemized in this report.

20.7.2 Qualifications for Special Inspectors. The same guidelines used for Special Inspections also discuss the minimum qualifications and certifications for various types of Special Inspectors. The qualifications for Special Inspectors are not code requirements, but provide guidance to the local building official. The minimum qualifications listed are from IAS AC291, *Accreditation Criteria for Special Inspection Agencies*, and are given as examples of qualifications. The IAS AC291 criterion undergoes periodic revisions to keep pace with code changes, industry standards, and changes in inspection protocol. The most current revision of IAS AC291 is available on the IAS website at www.iasonline.org/PDF/AC/AC291.pdf (see Table 1). Experience is hard to replace with education, and where the responsible professional for an agency provides his or her signature as evidence of competency for the special inspectors, it should be respected. Ultimately, the IBC places the responsibility for approval of special inspectors and special inspection agencies upon the building official. In many jurisdictions, the obtaining of the IAS certificate of accreditation is sufficient for proof of competency.

A. Experience

1. In order for experience to count toward qualifications, it must be based on verifiable work directly related to the category or type of inspection involved.

2. An engineering degree (BS), in addition to appropriate in-house training, may be substituted for not more than a year of experience. An engineering technology degree (AA), in addition to appropriate in-house training, may be substituted for not more than six months of experience. (Degree experience may not be substituted for more than half of the experience requirements in any category.)

3. Five or more years' experience as a qualified special inspector in one or more categories of work may fulfill up to half of the experience requirements in any category, at the discretion of the (agency's) responsible professional engineer.

B. Certification

Certification, when specified, is intended to mean successful completion of an ICC examination appropriate to the category of work involved.

C. Special Inspector in Training

1. The intent of the provision is to provide practical opportunities for an inspector to gain the needed experience to qualify as a special inspector.

2. An inspector who does not meet the qualifications for special inspector may be allowed to perform "special inspection" at the discretion of the agency's responsible professional engineer, provided one or more of the following conditions are met;

- Individual is working under the direct and continuous supervision of a special inspector fully qualified for the type of work involved.

- Individual is working under the indirect or periodic supervision of a special inspector and the scope of work is minor and/or routine and within the capabilities of the individual.

- Individual is specifically approved by the building official.

Table 20 gives the recommended qualifications for Special Inspectors for nearly every category of construction.

20.7.3. Special Inspection of Foundation and Earthwork. Many local authorities authorize a special inspector's engagement in soils inspection as well as foundation and earthwork inspection. Typically this inspector will be the geotechnical engineer or the testing laboratory in cooperation with the structural engineer. The building official should receive the special inspector's reports of the items with which the special inspector has been entrusted, including a statement that all items comply with requirements. Tables 20.2, 20.3, 20.4, and 20.5, taken from Tables 1705.3, 1705.6, 1705.7, and 1705.8 of the 2012 IBC, show the required scope of Special Inspection related to soils and foundations.

TABLE 20.1
MINIMUM QUALIFICATIONS FOR SPECIAL INSPECTORS
(TAKEN FROM ICC'S "MODEL PROGRAM FOR SPECIAL INSPECTION")

SPECIAL INSPECTION CATEGORY	REQUIRED EXPERIENCE	REQUIRED CERTIFICATION(S)	NOTES
Concrete construction (prestressed/precast)	Note c	ICC Prestressed SI and ICC Reinforced Concrete SI	
Reinforced concrete	Note c	ICC Reinforced Concrete SI or ACI Concrete Construction SI	
NDT	120 Hours for Level II	ANSI/ASNT-CP-189 NDT or SNT-TC-1a NDT	
Pier and pile foundations	Note c	NICET II (geotechnical or construction, or construction material testing or soils)	Note d
Post-installed structural anchors in concrete	Note c	ICC Reinforced Concrete SI or ACI Concrete Construction SI	
Soils	Note c	ICC Soils SI or NICET II (geotechnical or construction, or construction material testing or soils)	Note d
Spray-applied, fire-resistant materials Intumescent, fire-resistant coatings Mastic, fire-resistant coatings	Note c	ICC Spray-applied Fireproofing SI or ICC Fire Inspector	
Steel (high-strength bolting)	Note c	ICC Structural Steel and Bolting SI	
Steel (welding)	5 years minimum or in accordance with AWS	AWS CWI or ICC Structural Welding SI	
Masonry construction	Note c	ICC Structural Masonry SI	
Wood construction	Note c	ICC Commercial Building Inspector or ICC Residential Building Inspector	Note d
EIFS	Note c	AWCI EIFS Inspector	
Firestop systems	Note c	UL firestop examination or FM firestop examination	
Wall panels, curtain walls and veneers	Note c	ICC Commercial Building Inspector or ICC Residential Building Inspector	Note d
Smoke control systems	Note d	AABS technician certification	Note d
Mechanical systems	Note c	ICC Commercial Mechanical Inspector or ICC Residential Mechanical Inspector	

TABLE 20.1
MINIMUM QUALIFICATIONS FOR SPECIAL INSPECTORS
(TAKEN FROM ICC'S "MODEL PROGRAM FOR SPECIAL INSPECTION") (continued)

SPECIAL INSPECTION CATEGORY	REQUIRED EXPERIENCE	REQUIRED CERTIFICATION(S)	NOTES
Fuel-oil storage and piping systems	Note c	ICC Commercial Mechanical Inspector, ICC Residential Mechanical Inspector or API Above-ground Storage Tank Inspector	
Structural cold-formed steel	Note c	ICC Commercial Building Inspector or ICC Residential Building Inspector	Note d
Excavation-sheeting, shoring and bracing	Note c	NICET II (geotechnical or construction, or construction material testing or soils)	
High-pressure steam piping (welding)	5 years minimum or in accordance with AWS	AWS CWI or ICC Structural Welding SI	
Structural safety-stability and mechanical demolition	Note c	RDP, PE, BS engineering/architecture or valid site safety manager certification	
Site storm drainage disposal and detention	Note c	ICC Soils SI or NICET II (geotechnical or construction, or construction material testing or soils)	Note d
Sprinkler systems	Note c	ICC Commercial Building Inspector or ICC Residential Building Inspector	Note d
Standpipe systems	Note c	ICC Commercial Building Inspector or ICC Residential Building Inspector	Note d
Heating systems	Note c	ICC Commercial Mechanical Inspector or ICC Residential Mechanical Inspector	Note d
Chimneys	Note c	ICC Commercial Mechanical Inspector or ICC Residential Mechanical Inspector	Note d
Seismic isolation systems	Note c	RDP, PE, BS engineering/architecture	
Special Cases	Note c	ICC Commercial Building Inspector or ICC Residential Building Inspector	Note d

a. It is recognized that the development of qualified inspectors requires those individuals to obtain experience performing inspections of actual work. The requirements herein include such experience, as do some of the required certifications. To provide a vehicle for individuals to obtain this experience, they shall perform inspections in accordance with written associate or apprentice programs that are prepared by the SIA, approved by the IAS and meet the requirements of the local governing authority. These programs must include, at a minimum, passing certification exams, when available, administered by third-party agencies, such as ICC and ACI; in-house SIA and third-party training; observation by the associate or apprentice of inspections performed by certified inspectors; and performance by the associate or apprentice inspectors of duplicate inspections with certified inspectors. This written program will also define the use of associate or apprentice inspectors and will limit their use based upon the level of supervision and the complexity of the inspection assignment. The complexity of an assignment shall be minimal and will often be task specific. Supervision should be direct, with a certified inspector being present at the site with associate or apprentice. The associate or apprentice to certified inspector ratio on a project site shall not exceed 1:1. All documents related to work by an associate or apprentice inspector must be cosigned by a certified inspector. The written program must include documentation of compliance with the program.

b. When qualifications for special inspector are locally defined, by statute, ordinance or rule that meet or exceed the requirements outlined in this criteria, these local requirements shall be recognized.

c. Applicants shall comply with one of the following education and experience requirements:

    1. PE, licensed architects or Registered Design Professional, and a minimum of three months of relevant work experience; or

    2. BS in engineering, architecture or physical science, and a minimum of six months of relevant work experience; or

    3. Two years of verified college or technical school (a copy of a diploma or transcript required), and a minimum of one year of relevant work experience; or

    4. High school or equivalent graduate, and a minimum of two years of verified relevant work experience; or

    5. A minimum of three years of verified relevant work experience.

d. PE, licensed architects or Registered Design Professional are exempt from the required certifications listed in this table, but are subject to on-site assessment of competence by IAS.

TABLE 20.2
REQUIRED VERIFICATION AND INSPECTION OF CONCRETE CONSTRUCTION
(TAKEN FROM IBC 2012 TABLE 1705.3)

VERIFICATION AND INSPECTION	CONTINUOUS	PERIODIC	REFERENCED STANDARD[a]	IBC REFERENCE
1. Inspection of reinforcing steel, including prestressing tendons, and placement.	—	X	ACI 318: 3.5, 7.1-7.7	1910.4
2. Inspection of reinforcing steel welding in accordance with Table 1705.2.2, Item 2b.	—	—	AWS D1.4 ACI 318: 3.5.2	—
3. Inspection of anchors cast in concrete where allowable loads have been increased or where strength design is used.	—	X	ACI 318: 8.1.3, 21.2.8	1908.5, 1909.1
4. Inspection of anchors post-installed in hardened concrete members[b].	—	X	ACI 318: 3.8.6, 8.1.3, 21.2.8	1909.1
5. Verifying use of required design mix.	—	X	ACI 318: Ch. 4, 5.2-5.4	1904.2, 1910.2, 1910.3
6. At the time fresh concrete is sampled to fabricate specimens for strength tests, perform slump and air content tests, and determine the temperature of the concrete.	X	—	ASTM C 172 ASTM C 31 ACI 318: 5.6, 5.8	1910.10
7. Inspection of concrete and shotcrete placement for proper application techniques.	X	—	ACI 318: 5.9, 5.10	1910.6, 1910.7, 1910.8
8. Inspection for maintenance of specified curing temperature and techniques.	—	X	ACI 318: 5.11-5.13	1910.9
9. Inspection of prestressed concrete: a. Application of prestressing forces. b. Grouting of bonded prestressing tendons in the seismic force-resisting system.	X X	—	ACI 318: 18.20 ACI 318: 18.18.4	—
10. Erection of precast concrete members.	—	X	ACI 318: Ch. 16	—
11. Verification of in-situ concrete strength, prior to stressing of tendons in post-tensioned concrete and prior to removal of shores and forms from beams and structural slabs.	—	X	ACI 318: 6.2	—
12. Inspect formwork for shape, location and dimensions of the concrete member being formed.	—	X	ACI 318: 6.1.1	—

For SI: 1 inch = 25.4 mm.

a. Where applicable, see also Section 1705.11, Special inspections for seismic resistance.

b. Specific requirements for special inspection shall be included in the research report for the anchor issued by an approved source in accordance with ACI 355.2 or other qualification procedures. Where specific requirements are not provided, special inspection requirements shall be specified by the registered design professional and shall be approved by the building official prior to the commencement of the work.

TABLE 20.3
REQUIRED VERIFICATION AND INSPECTION OF SOILS
(TAKEN FROM TABLE 1705.6 OF THE 2012 IBC)

VERIFICATION AND INSPECTION TASK	CONTINUOUS DURING TASK LISTED	PERIODICALLY DURING TASK LISTED
1. Verify materials below shallow foundations are adequate to achieve the design bearing capacity.	—	X
2. Verify excavations are extended to proper depth and have reached proper material.	—	X
3. Perform classification and testing of compacted fill materials.	—	X
4. Verify use of proper materials, densities and lift thicknesses during placement and compaction of compacted fill.	X	—
5. Prior to placement of compacted fill, observe subgrade and verify that site has been prepared properly.	—	X

TABLE 20.4
REQUIRED VERIFICATION AND INSPECTION OF PILE FOUNDATIONS
(TAKEN FROM TABLE 1705.7 OF THE 2012 IBC)

VERIFICATION AND INSPECTION TASK	CONTINUOUS DURING TASK LISTED	PERIODICALLY DURING TASK LISTED
1. Verify element materials, sizes and lengths comply with the requirements.	X	—
2. Determine capacities of test elements and conduct additional load tests, as required.	X	—
3. Observe driving operations and maintain complete and accurate records for each element.	X	—
4. Verify placement locations and plumbness, confirm type and size of hammer, record number of blows per foot of penetration, determine required penetrations to achieve design capacity, record tip and butt elevations and document any damage to foundation element.	X	—
5. For steel elements, perform additional inspections in accordance with Section 1705.2.	—	—
6. For concrete elements and concrete-filled elements, perform additional inspections in accordance with Section 1705.3.	—	—
7. For specialty elements, perform additional inspections as determined by the registered design professional in responsible charge.	—	—

TABLE 20.5
REQUIRED VERIFICATION AND INSPECTION OF PIER FOUNDATIONS
(TAKEN FROM TABLE 1705.8 OF THE 2012 IBC)

VERIFICATION AND INSPECTION TASK	CONTINUOUS DURING TASK LISTED	PERIODICALLY DURING TASK LISTED
1. Observe drilling operations and maintain complete and accurate records for each element.	X	—
2. Verify placement locations and plumbness, confirm element diameters, bell diameters (if applicable), lengths, embedment into bedrock (if applicable) and adequate end-bearing strata capacity. Record concrete or grout volumes.	X	—
3. For concrete elements, perform additional inspections in accordance with Section 1705.3.	—	—

20.1 TEST QUESTIONS

MULTIPLE CHOICE

1. Inspection of construction of foundations (select one):
   a. is not important if the foundation plans are properly prepared
   b. is always important
   c. is not required since contractors are experienced
   d. is too dangerous for most inspectors

2. Written inspection reports on foundation or earthwork construction (select one):
   a. are a waste of time unless concrete placement is involved
   b. are very important
   c. do not need to be detailed
   d. are not required since oral reports are faster

3. Preparing a construction site to drain off rain water (select one)
   a. is only an inconvenience to the contractor and not necessary
   b. is important for a good job
   c. is not necessary if rain is not imminent

4. The proper preparation of the site for foundation construction (select one):
   a. is not critical to the proper performance of the foundation
   b. is as important as confirming good concrete work
   c. is not usually needed

5. During concrete placement, the inspector should obtain (select one):
   a. the concrete plant batch mix tickets
   b. the ready mix truck driver's license information
   c. date of last truck mechanical inspection
   d. delivery route traffic conditions

6. The location of structural elements such as walls or plinths resting on a spread or strip footing should be checked to (select one):
   a. ensure that the plinth locations fall on the footing
   b. ensure that eccentricity is not introduced in the foundation element
   c. verify that any offset is toward the prevailing wind direction

7. Placing concrete vibration (select one):
   a. is not necessary since concrete will settle under its own weight
   b. can damage fresh concrete and should not be permitted
   c. is important to eliminate voids
   d. should not be used since reinforcing steel can be displaced

8. Monolithic concrete placement involves (select one):
   a. filling the entire form
   b. nearly continuous placement of concrete against already placed fresh concrete
   c. placement against a locally prominent geologic rock formation

9. If groundwater or caving conditions are expected, the pier contractor should have available the proper (select one)
   a. shovels
   b. rebar
   c. driving hammer
   d. casing
   e. diesel fuel

10. Minimum information the inspector should record during pile driving includes the following (select three):
    a. the type of piling in accordance with plans
    b. length of leads
    c. depths of establishment
    d. driving energy
    e. piling age
    f. available safety equipment

11. A typical $1/2$-inch, high-strength steel-strand, post-tensioning tendon is stressed to a residual after relaxation of about _____ pounds after the concrete hardens (select one):
    a. 10,000
    b. 100,000
    c. 26,600
    d. 22,400

12. During inspection of post-tensioned foundations, two things that should be recorded for each tendon are (select two):
    a. azimuth direction
    b. final stressing force load
    c. date delivered
    d. elongation
    e. width of form

13. Unless noted otherwise on the engineer's plans, earth slopes to shed water away from foundations are typically a minimum of_____ inches fall over 10 feet horizontally (select one):
    a. 24
    b. 36
    c. 2
    d. 6

Chapter 21

FOUNDATION NON-PERFORMANCE

Foundation non-performance

Chapter 21

FOUNDATION NON-PERFORMANCE

21.1 Foundation Non-performance

After construction, a foundation may not be performing as expected, either due to improper geotechnical information, improper design, poor construction, or all of these. Such conditions require evaluation by experts as to whether remediation is necessary and what type of remediation should be utilized. If a foundation does not perform properly, the entire structure will be affected, with the results ranging from minor cosmetic distress to major structural failure. Non-performance may be related to bearing failure, excessive settlement, slope instability, or heave due to expansive clays. Inadequate bearing capacity under footings can be a serious problem. Settlement can occur because of improper compaction of fills or because the native soil is compressible and has begun to consolidate. Slope instability can be gradual downhill creep, which will continually add to the distress of the building over time, or it can be major instability in which the situation is obvious. Heave of expansive clays may occur due to natural environmental conditions or because of the concentration of water due to poor drainage or leaking plumbing pipes. Sometimes subsurface water may be feeding the expansive clays in an unexpected manner.

21.2 Evidence of Non-performance

Foundation non-performance is usually recognized because of distress in the walls, ceiling, or doors of the superstructure. Non-performance could be excessive tilt of the entire foundation, or it could be distortional movement of the foundation. Tilt will not necessarily cause distress in the superstructure unless it becomes too great. If the tilt of a structure becomes greater than about one percent for a one- or two-story building, the center of gravity of the walls may go beyond the contact area of the wall-bearing surfaces, and it is possible for brick veneers to fall or downhill racking of the superstructure to occur. Figure 21.1 illustrates tilt.

Distortion of the foundation could be caused by various footings settling while others are not, or entire segments of the building may be settling if, for example, half the building is on fill and is settling and the other half is not. Expansive clays typically cause distortion of the foundation and may also cause tilt. If distortion exceeds a certain amount of foundation curvature, the exterior and interior wall finishes will crack, the door frames will rack out of square (and the doors will not operate properly), and underground utilities may be ruptured. Figure 21.2 and 21.3 show two modes of foundation distortion.

Figures 21.4 through 21.7 illustrate various manifestations of foundation non-performance.

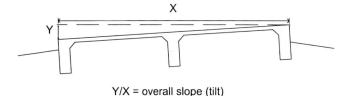

Y/X = overall slope (tilt)

FIGURE 21.1
OVERALL TILT OF FOUNDATION

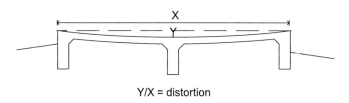

Y/X = distortion

FIGURE 21.2
DISTORTION OF FOUNDATION – EDGE-LIFT MODE

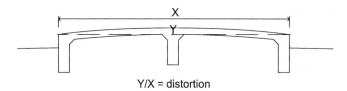

Y/X = distortion

FIGURE 21.3
DISTORTION OF FOUNDATION – EDGE DROP OR CENTER-LIFT MODE

21.3 Remediation Design

Design of foundation remediation requires the services of an expert familiar with geotechnical as well as structural engineering. Often a foundation repair contractor will hold himself or herself out as an expert in this subject, and some may truly be. However, many of these contractors are not well versed in the causes and remediation procedures and may simply be interested in selling a remediation job to an owner.

Typical remediation for foundations that have settled or heaved differentially includes the use of supporting piers or piles in various locations under the foundation and re-leveling the structure on these elements. Other approaches include moisture barriers and drains around the perimeter of the structure to stop subsurface water from penetrating, either due to seepage water in the ground or natural environmental movement of moisture in and out. Sometimes chemical or grout injection beneath the footings can stabilize the situation. In all cases, the engineer must make a decision as to whether to re-level the foundation or simply hold it in place and repair the superstructure.

FIGURE 21.4
FOUNDATION NON-PERFORMANCE – BRICK CRACKING

FIGURE 21.5
FOUNDATION NON-PERFORMANCE – DISPLACED BRICK VENEER. THIS IS A DANGEROUS SITUATION DUE TO FALLING BRICK.

In cases of settlement due to improperly compacted fill, often foundation remediation can be permanent and effective. If long-term, deep-seated settlement due to consolidating natural soils is the cause, the phenomenon may continue even after remediation. The same can be said for expansive clay soils, and sometimes it is impossible to totally stop the movement, although a good remediation job may greatly reduce the effects of the movement. If expansive clays are causing the problem, placing piers under the foundation can only prevent parts of the foundation from going down, but the piers will not stop the foundation from going up should the clay continue to heave. It is virtually impossible to put a deep pier or pile down and securely attach it to a soil-supported slab foundation with the hope that it will not heave up. The pressures of expanding clay when confined can be immense, and such approaches are almost always doomed to failure.

FIGURE 21.6
FOUNDATION NON-PERFORMANCE – WALL CRACK AND DOOR FRAME OUT OF SQUARE. IF DOORS CANNOT BE OPENED OR LOCKED, THEY ARE SAFETY HAZARDS.

FIGURE 21.7
FOUNDATION NON-PERFORMANCE – HOUSE IS BENT. SEE ROOF LINE. CAUSE IS EXPANSIVE CLAY AND LEAKING PLUMBING WHERE GRASS IS GREENER.

Figures 21.8 and 21.9 illustrate the installation of piers for foundation remediation.

21.4 Remediation Construction Inspection

The same principles hold for this type of inspection as for any other. The remediation plan should be prepared by an engineer, and preferably that engineer will provide inspection services to ensure that his or her plan is properly implemented. Since foundation remediation is often carried out under the foundation, either through tunnels or pits, it is frequently inconvenient and difficult to properly inspect the work that is being done, and contractors are not going to volunteer that they did not do their job. Therefore, inspection by someone other than the contractor is necessary.

Piers or piles may be a part of the remediation program. The inspection process should include a listing of each pier or pile with the depths of establishment and ground conditions observed. The establishment depth should be compared with the design depths and may be modified in the field with the design engineer's approval due to conditions encountered as described previously under pier or piling inspection.

If water diversion is part of the remediation plan, careful inspection is necessary to ensure that surface or subsurface drains are constructed as described previously and will flow downhill continuously. Subsurface drains must be constructed with no bellies or reservoirs in the drainage material, which could hold water instead of carrying it away.

If a moisture barrier is part of the remediation, care should be taken that the barrier is properly applied and no gaps are permitted. Care should also be taken if the barrier is to be adhered to the exterior grade beam. The inspector should see that a good bond is achieved so that the barrier does not peel off later and permit rain water to sheet down inside the barrier. It is a good plan to have a nailing strip or reglet at the top of the barrier where it connects to the concrete to avoid this situation. The barrier should be below the surface of the ground to protect it from ultraviolet light, and it should be composed of material stout enough to avoid random damage. The seal of the barrier around utility pipes or buried cable penetrations is especially important for a successful job.

FIGURE 21.8
INSTALLING REMEDIAL PIERS AROUND EXTERIOR OF BUILDING

FIGURE 21.9
INSTALLING INTERIOR REMEDIAL PIER BY OPENING FOUNDATION. BECAUSE OF THE INTERIOR DESTRUCTION, INTERIOR PIERS ARE OFTEN INSTALLED BY TUNNELING ACCESS BELOW THE FOUNDATION.

21.5 TEST QUESTIONS

MULTIPLE CHOICE

1. If a foundation has moved with a tilt exceeding _____ percent, it is cause for concern (select one):
    a. 10
    b. 0.5
    c. .25
    d. 1

2. A technique typically used for used for remediation of foundations that have settled or heaved differentially is (select one):
    a. adding supporting piers or piles
    b. running water pipes in the attic
    c. blowing hot air below the foundation

3. If expansive clay soils are causing foundation non-performance, placing piers under the foundation helps prevent all parts of the foundation from (select one):
    a. going up or down
    b. going down
    c. going up

4. Inspection of remediation construction (select one):
    a. it is not necessary since it is secondary construction
    b. should not be performed due to liability issues
    c. is important

5. Foundation non-performance may be related to (select three):
    a. lightning strikes
    b. fill settlement
    c. high winds
    d. expansive clay
    e. downhill creep
    f. piers too deep

6. Heave of expansive clays may occur due to natural environmental conditions, or because of (select one):
    a. the concentration of water due to poor drainage
    b. lack of lime
    c. sudden drop in temperature
    d. buried electrical lines

7. Name two places in a building structure where foundation non-performance is usually first recognized (select two):
    a. distress in roof covering
    b. dry wall cracking
    c. door fit
    d. drying grass
    e. clogged plumbing

GLOSSARY

ACI: American Concrete Institute

aggregate: coarse material generally graded to some specification used in concrete, hot mix for paving and for select fill.

aggressive soils: soils that because of high acidity or high alkalinity can attack the materials of construction such as concrete and steel.

alkaline: the property of a material to have a pH greater than 7.0. Also known as basic. An example would be quick line.

alluvium: soil material of gravel to silt and clay sizes deposited by moving water, typically found overlying older geologic formations.

ASCE: American Society of Civil Engineers

ASCE 7: Minimum Design Loads for Buildings and Other Structures

anchored steel bulkhead: a retaining structure typically used at water fronts, consisting of driven sheet piling with tiebacks connected to deadmen beyond the zone of potential movement.

Appendix J: the appendix in the 2006 and 2012 IBC that is devoted to excavation and grading.

aquifers: underground formations which, because of porosity, can contain and transport groundwater.

architect: a design professional normally responsible for the overall design of a building project. Architects do not generally design engineering works such as dams, bridges and treatment plants.

ASTM: American Society for Testing and Materials

at rest: the state of retained earth mass behind a retaining structure which exerts higher lateral pressures because it is not permitted to move laterally, thus mobilizing internal shear strength.

Atterberg Limits: these are soil index properties determined in the laboratory which define many characteristics of fine grained soils. They include the Plastic Limit, Liquid Limit, Plasticity Index and Shrinkage Limit.

backfill: fill placed behind retaining walls, basements or within utility trenches. The word backfill implies that a previous excavation was done and that the fill is being placed "back". Backfill may also be a specified, imported material called "select fill".

bearing: as used in foundation construction this refers to the support capacity of the soil or rock on which the footing elements are established.

bedrock: this is rock which is intact and in the position in which it was originally formed. Loose pieces within a slope or fill such as boulders are not bedrock.

bell: the enlarged base of a drilled pier produced by under-reaming. The bell is used to enlarge the end bearing area of a drilled pier.

bench: a bench is a level area, either excavated or naturally occurring on a sloping site.

berm: a berm is an earth-constructed low levee, which is intended to divert surface water, provide a landscape feature, or provide noise protection.

bluff: a bluff is a geologic condition synonymous with a cliff. A bluff is frequently formed by either large-scale faulting or erosion on a large scale, which leaves a nearly vertical face of rock or earth.

boring log: a written record with symbols and words which depicts the materials and depths encountered during exploratory drilling for a site investigation.

borrow: borrow is earth used for the purposes of fill or special construction, originating off site or somewhere else on the construction site. In many parts of the country rural road sections refer to a "bar ditch," which is the ditch on each side of the roadway from which the roadway embankment soil was excavated, and which then serve as drainage ditches. The term "bar ditch" is a colloquialism for "borrow ditch."

building official: the person who is legally responsible for enforcement of the adopted building code in a particular locality.

buoyancy: the property of submerged material to weigh less than if it was not submerged. The effective unit weight of s submerged soil material is the non-submerged unit weight minus the unit weight of water.

cantilever wall: this refers to a retaining structure typically constructed out of reinforced concrete, which has a horizontal footing and a vertical wall that is securely attached to resist lateral earth forces.

capillary: the capacity of a soil, usually fine sand to silt and clays, to "wick up" water above the normal water table.

casing: in installation of drilled piers a temporary steel cylinder placed in the hole to keep the hole from caving in unstable ground, permitting placement of concrete and steel to construct the pier. Casing may be permanently left in place or considered temporary and withdrawn as the concrete is placed.

cementation: when used with regard to soils this indicates the soil material has been bound together by chemical bonds such as lime, producing a stronger soil condition.

chairs: used in reinforced concrete technology these are small metal or plastic items placed in the forms to position the reinforcing steel. They are left in place after the concrete is installed.

checker boarding: this refers to constructing remediation in widely spaced parts so is not to cause the complete collapse of the slope or building being worked upon. After the checker boarded initial construction further construction is checker boarded in between until finally the entire remediation project is complete.

chemical and water injection: injecting water borne chemicals or plain water into the ground under pressure to depths of five to ten feet in order to change the properties of the soil. Used in stabilization of sites, which may lack sufficient strength or may be an expansive clay.

clay: grain sizes of a soil mixture generally referred to as that smaller than .002 mm.

CLSM: Controlled Low Strength Material usually composed of a mixture of Portland cement, fly ash, air entrainment, water and sand. Its purpose is to provide backfill that does not need to be mechanically compacted, but will gain strength rapidly. CLSM in not intended to become so strong that it is inconvenient to remove it later, such as for utility maintenance.

cobbles: soils materials that have individual particle sizes ranging from 3 inches to 12 inches.

coefficient of friction: the value to be multiplied by the contact or "normal" force pressing to another surface to yield the amount of sliding friction that can be generated.

cold joint: a joint in concrete at a planned or unplanned interruption of concrete placement. If not installed according to a plan detail, it will be a concrete defect.

colluvium: soil materials which may be found as a natural collapse of a bluff or slide which has overrun and covered other materials.

compaction: the increase in density of a soil material usually by mechanical means. The earth is packed together tightly to expel the air voids, producing a heavier unit weight, higher strength, and stability. Compaction could also occur in nature as a result of overburden compressing lower soils, expelling water over an extended time period. In natural deposits this is more likely to be called consolidation.

compressible: the property of a soil to be reduced in volume by squeezing out air or water from its void structure by construction loadings. A compressible soil may result in unacceptable settlement of structures.

concrete: a mixture of gravel, sand, Portland cement, water and sometimes chemical additives to produce a plastic mixture which can be placed in forms and will harden to a rock-like consistency. Widely used in construction.

consolidometer: laboratory equipment which is used to measure the compressibility of a soil.

constructability: considerations of the factors that may come into play in determining whether or not a designed structure can be built efficiently.

contour: a line on a topographic map of equal elevation. The closer together such lines are the steeper the slope and vice versa.

core: a cylindrical sample typically of rock extracted from the ground using a core barrel during exploratory drilling.

CR: core recovery. The length of a recovered core is divided by the length of the core run and expressed as a percentage.

crawl space: the area beneath the first floor of the structure, which has sufficient separation from the ground so that a person could crawl through it.

creep: a very slow movement down slope of near-surface soil and loose rocks, which will eventually produce damage to constructed facilities located on the surface. Creep is usually seasonally cyclic, moving more in wet periods and slowing or stopping during dry periods. Creep is also seen in snow packs on hillsides and, in both snow and in earth, if creep becomes significantly large, it will herald an avalanche or landslide.

crib wall: a retaining structure in a box-like configuration composed of timbers or pre-cast concrete and typically filled with gravel or crushed stone to provide resistance to lateral movement of soil. A gravity retaining structure.

curvature: when used with regard to constructed facilities this indicates the deviation from a straight line or flat plane of the floor or other segments of the structure. Curvature will result in cracking of walls and other parts of the structure.

deep foundations: established below the immediate surface of the ground utilizing procedures such as drilled piers or piling.

distress: when used in terms of structures this refers to manifestations of movement creating visible damage. Distress may indicate a significant structural problem or may be only cosmetic, affecting appearance.

down drag: the condition in which disturbed soil may settle around a pile or a pier and create downward force on the foundation element in excess of what was assumed for normal dead and live loads.

drainage: the process of removing either surface or subsurface water and channeling it in a way so as not to be a problem to the constructed work.

drilled piers: deep foundation elements placed by cutting a cylindrical hole with an auger and removing the earth to create an earth form for placing of concrete and steel to create a pier.

driving log: the record of number of blows per foot of a pile driving rig recorded against depth.

dry well: a hole in the ground typically filled with coarse gravel with the hope that drainage can be directed into this hole. A dry well seldom is effective for subsurface or surface drainage disposal.

earth pressures: pressures generated by the earth against a retaining wall or basement wall as opposed to pressures imposed on the earth by a footing.

earthquake: the shaking of the earth, sometimes violent, usually resulting from slippage of a fault underground.

earthwork: the process of moving earth either by cutting or filling to prepare a site to receive the intended construction.

easement: a land use legal document usually filed with the local county which provides a right for a utility or other entities to do certain types of construction or installation. Examples would be easements for drainage, water, sewer or electrical lines. An easement is different from a Right-of-Way in that the property owner can still make limited use of the land, but the public authorities have the right to construct and maintain public works.

eccentricity: the placing of a load onto an item, such as a footing, off center, which would produce tilting or turning action as opposed to a centered load.

engineer: the design professional whose responsibilities may include directing the site investigation, preparing structural, mechanical, electrical or plumbing designs with regard to building or other structures. Engineers are qualified in a particular discipline based on specialized education, typically four years of qualifying experience and passing of rigorous exams. The state government typically licenses engineers to practice within their jurisdiction.

equivalent fluid: a method used in the design of retaining structures to represent the earth as a fluid load behind the wall with different unit weights depending on the characteristics of the soil.

erosion: erosion involves the loss of soil or rock material due to the actions of water, ice, or wind. Erosion can also be applied to constructed works, such as concrete structures, if sufficient erosional forces have been at work.

expansive: property of a soil material to expand or increase volume when permitted access to water. Expansive soils also have the property of reducing volume when drying.

factor of safety: this is the amount of over design built in to structures to allow for unknowns in the design assumptions, calculations or construction.

fault: the rupture of the earth or rocks due to local or continental plate movements and stress. Sudden movement along a fault will result in an earthquake.

field expedient: procedures used in soils engineering to estimate properties of soils without the benefit of full laboratory.

fill: any soil material that has not been placed by nature.

fill lift: when compacting fills, it is usual to place the uncompacted material in layers or "lifts." These layers are typically 9 to 12 inches thick prior to compaction and will be reduced in thickness by about $1/3$ after compaction. The use of thin lifts is important in compacting soil to ensure that the compacting forces are able to operate on the full depth of loose soil to provide the required densification.

flow line: a term used in drainage or utilities. When used with regard to relatively small diameter pipes the flow line of a gravity sewer or storm drain is generally considered the inside bottom of the pipe.

footings: the structural element which transfers the load of the super structure to the earth below.

french drain: also known as a subsurface drain, this item is used to remove water from below the surface of the ground and carry it away in a controlled fashion.

γh: as used with regard to unsaturated soil mechanics, this term is the expansion-compression index of a soil and is used to calculate the amount of volume change that would take place when the suction or energy levels within the soil change.

geo-hazards: a condition of the earth which could cause a safety concern or damage to a structure.

geologic age: a division based on time which geologists use to separate the various formations that make up the earth. Geologic Ages range from nearly the beginning of the earth to the present time.

geology: geology is the science of the origin, formation and description of materials of the earth.

geo-structural engineering: a term used to describe the engineering design of any structure that is involved with the earth, such as foundations, retaining walls, paving and stabilization of slopes. It is a combination of the disciplines of geotechnical and structural engineering.

geotechnical engineering: is a term which involves the knowledge of applied geology, soil formation and composition, reaction of soil to loadings and strength and stability capacity of soils. Geotechnical engineering also is concerned with ways of obtaining the geotechnical information from the field or laboratory and in performing analysis of such information for the purpose of guiding the design of the foundations or other structures to be placed against the earth.

geotextile: this is a manufactured product which may be perforated or woven plastic or other similar material that is used to permit passage of water, but not soil particles.

GPS: Global Positioning System, which is a satellite based navigation system consisting of a network of 24 orbiting satellites. Using GPS one can determine the location of a point on the ground, sometimes with great precision, depending on the sophistication of the equipment.

grade: grade is a term commonly used to describe either the initial or final elevation of the ground surface. The grade prior to construction would be called existing grade or original grade, and the finished grade would be the desired final elevations of the site in accordance with the plans, both represented by contours or spot elevations on topographical maps. The term "grading," refers to the processes of getting a site configured to the grades required by the plans.

hammer: when used with regard to driving piles this is the piece of equipment that provides the impact energy to the top of the pile to force it into the earth. Pile driving hammers come in a variety of configurations including a gravity drop weight, steam hammers, compressed air hammers, hydraulic hammers and vibratory hammers.

HMAC: hot mixed asphaltic concrete. Frequently used as a surface course for flexible pavement.

hollow stem: with regard to site investigations this is a type of continuous flight auger which has an open or hollow interior, which permits sample tools to be penetrated into the ground as the auger is advanced. This is typically used in caving hole conditions. In effect the boring is advanced utilizing its own casing.

honeycomb: with regard to concrete technology this is the result of poor concrete placement technique or lack of vibration which permits extensive voids to appear in the hardened concrete.

hydrocollapse: the property of certain type of soils, typically found in semi-arid or arid regions which support an open void structure by means of cemented contacts between particles. Upon saturation the cementation is softened and the soil profile may collapse, sometimes several feet.

hydroconsolidation: this is the property of certain soil fills to collapse because void spaces which were held open by arching of hard clods were later softened by water. The collapse can be sudden and often can be several feet.

hydrodynamic settlement: this is a term used to describe the settlement of a foundation of a structure when placed on a soft saturated soil with the water being forced out by the pressure applied by the construction.

hydrometer test: a laboratory test used to define the distribution of grain sizes smaller than the #200 Sieve.

hydrostatic: the pressure of at-rest water, becoming greater with depth in a linear fashion.

hydraulic mulch: a sprayed mixture of water, fertilizer, mulching materials, and seeds applied to provide a quick grass start and protection of newly graded raw earth slopes. Some building departments require hydraulic mulch to be in place or even grass growing on raw earth slopes prior to the final release of the building or site permit.

IBC: *International Building Code*

igneous: rock formations formed by the hardening of molten lava or magma.

index properties: soil properties resulting from lab testing which can be performed rapidly and inexpensively, which are related to engineering properties.

inspection: also known as project observation, this is the process of noting whether or not the construction is in accordance with approved plans and specifications.

IRC: *International Residential Code*

kelly bar: a square steel bar in a pier drilling rig that can move up or down and transmits the rotary motion to the drilling tools such as augers or belling tool.

kip: 1,000 pounds

ksf: kips per square foot

laboratory dry density: most commonly used as the maximum laboratory dry density obtained from the Proctor test in the laboratory. This is the maximum dry unit weight to which a soil can be compacted in the laboratory under the specified compaction procedure.

landslides: also known as landslips, this is a dramatic movement of a slope of earth and rocks in a downward and outward direction. Sometimes the movement is relatively slow on the order of a few feet a week, but sometimes it is rapid, moving hundreds of feet within a few hours.

limestone: a rock material formed by deposition, usually in marine or freshwater lake environments involving precipitation of calcium carbonate, either by chemical precipitation or the shells of marine organisms. This is a sedimentary type rock.

liquifaction: the condition which results when a saturated soil is shaken by an earthquake, causing it to lose all internal shear strength and become liquid.

Loess: a wind borne deposit which has become cemented with an open void structure. Loess is stable, even standing on vertical cuts for an indefinite period of time. However, when saturated it may suddenly lose strength and collapse.

LRFD: Load and Resistance Factor Design. In this structural analysis procedure the factor of safety is applied to the loadings and all the materials and elements of construction are analyzed at their ultimate stress capacity.

masonry: a construction technique using either rough or carefully cut stone or brick which may be mortared or non-mortared to form walls or other elements of a structure.

mat: a foundation type which typically consists of a plate of reinforced concrete with sufficient strength and stiffness to distribute the loads of the structure in a uniform manner to the soil.

metamorphic: a type of rock which can be formed either from igneous or sedimentary rocks by the application of heat and pressure over a long period of time.

mph: miles per hour. Used to express the speed of vehicles.

mudflow: a river of soil, water, rock, and other debris. It is too soft to walk on and will rapidly flow as a fluid, moving from higher elevations to lower, and it can be a nearly unstoppable force of destruction. Mudflows are triggered by heavy rain, usually in mountainous or hilly areas. The runoff picks up a considerable erosional load, often involving freshly placed earth from construction sites.

nuclear density testing: the procedure by which testing laboratories determine the field dry density and the field moisture content of soil or paving by means of probes which have radioactive transmitters and sensors. The sensors detect the transmission of the radiation or the product of radiation. Nuclear density devices must be licensed, typically by the State Health Department and have to be operated by qualified individuals because of safety concerns.

optimum water content: this refers to the water content required to achieve the maximum laboratory density in any of the Proctor test procedures. From a dry state a soil will compact to higher and higher densities as more water is mixed with the soil until it reaches the optimum water content. Addition of more water past this point will cause the soil to become saturated and lower dry densities will be achieved. The optimum water content in the field will permit maximum compaction efficiency.

organics: plant or vegetable matter found in soil. This can be found in man made landfills or in nature due to alluvium deposits or in, for example, "peat bogs", which are nearly

100% organic. Organics generally pose a major problem for foundation construction due to their capacity to produce large settlements.

pcf: pounds per cubic foot.

permeability: property of a soil or other material to transmit water. The speed and quantity of water which will transmit through the material is based on the permeability properties of the material, but also depends on the driving head.

permeameter: a laboratory device to test for permeability

pF: a term used in unsaturated soil mechanics which refers to the level of suction or negative energy present in a soil mass. It is defined as the logarithm to the Base 10 of a column of water with a given height in centimeters.

pH: a term used in chemistry to denote the acidity or alkalinity of a substance. A pH below 7.0 indicates an increasingly acidic condition and above 7.0 up to 14 is an increasingly alkaline condition. The further away the pH is from 7.0 in either direction the more active the chemical is when attacking other substances.

π: π is a mathematical term that relates the diameter of a circle to the circumference or outer perimeter of the circle. The relationship for engineering purposes typically is taken as 3.1416.

PI: the abbreviation for Plasticity Index, one of the Atterberg Limits determinations. It is obtained by subtracting the Plastic Limit from the Liquid Limit.

Pile: a linear member driven vertically into the ground to reach areas of higher bearing capacity or to develop skin friction along its length to produce adequate support for the loads from the structure above it. This is a type of deep foundation.

plant mix ticket: the print out of the weights of various materials going into making up a particular batch of concrete in a ready mix truck. The plant mix ticket should list the pounds of cement, the pounds of gravel and sand and the weight or quantity of water, as well as any chemical add-mixtures. It may also be called a batch ticket.

plasticity: the property of a solid to be deformed under pressure and retain that shape without recovery or disruption.

pocket penetrometer: a spring loaded device which presses a cylindrical steel rod into the soil a given distance, calibrated to provide an estimate of the unconfined compressive strength of the soil, typically in tons per square foot.

post-tensioning: a method for constructing foundations or other parts of a structure in which high strength steel cables are placed within the concrete in a greased sheath permitting the post-tensioning tendon to be pulled from the anchorages at the outside faces to compress the concrete.

pre-stress: this refers to compressive stress usually in pounds per square inch applied across the concrete section by the post-tensioning or pre-tensioning procedures to reinforce concrete.

proctor: there are two widely used specifications for producing a Proctor curve in the laboratory, also known as the moisture-density relationship. One is the Standard Proctor which is described in ASTM D-698. The other is the Modified Proctor procedure described in ASTM D-1557. These procedures are named for a man named Proctor who developed them in the 1930's. Densities obtained in the field are compared to those obtained in the laboratory under the specified type of compaction specification and expressed as a percentage. Typically specifications call for 90, 95 or 100% of the maximum laboratory dry density to be obtained in the field. Failure to obtain these percentages would cause the contractor to have to re-work the fill and repeat the field compaction procedures.

proof rolling: a method used in earth work to verify that adequate compaction and stability has been obtained in a uniform manner on a fill or a subgrade. It is used most extensively in pavement construction (but also for building site fill) to uniformly test the area which would not be fully tested by widely spaced field density tests.

psf: pounds per square foot.

psi: pounds per square inch.

PTI: Post-Tensioning Institute

PVC: Polyvinyl Chloride which is a plastic material used in construction in pipes and vapor barrier sheets.

raft: another term for a foundation mat.

reinforcement: used in conjunction with concrete construction, this refers to steel rods, which are deformed to increase the bonding with the concrete to provide tensile capacity to the concrete section. Also known as rebar or reinforcing steel.

reveal: used in conjunction with soil supported slab foundations, indicates the edge exposed above the soil up to the level of the finished floor of the structure.

reverse skin friction: this is the result of expanding clay soil which tends to drag a pier or pile out of the ground as it expands upward.

right-of-way: the strip of land typically dedicated to public use for constructing roads or major utility lines.

rip rap: placing of stone or thin concrete on a sloped earth surface to protect against erosion. It is often used to provide protection from breaking waves or running water on shore-lines.

rock: a natural material of practically irreversibly cemented minerals and is generally stronger than very strong soils.

rotary bit: used in conjunction with drilling test holes, water wells or oil wells, this is a bit usually with hardened ridges or points, which is rotated by the drill pipe and penetrates into the earth or rock by removing pieces or chips as it progresses downward. Typically water or air is used to remove the cuttings to the surface.

rough grading: the general shaping of the earth of a site prior to initiation of foundation, pavement, or utility construction.

RQD: Rock Quality Designation. All pieces of a recovered core run less than four inches are removed and the remaining length of core is divided by the length of the core run and expressed as a percentage.

sampler: a device used in site investigation work to obtain samples of the soils from various depths for further examination or testing in the laboratory.

sand: a soil material which can range in size from about $^{1}/_{4}"$ down to that which is retained on the #200 sieve. Sand is considered a coarse grain soil material.

sedimentary: in geologic terms these are rocks formed by deposits, typically in water and then cemented through pressure over extended periods of time.

segmental concrete walls: retaining structures constructed using pre-cast units of concrete often formed to look much like natural stone. The segmental units are connected into the embankment behind it by tensile plastic reinforcement layers known as geo-grids. Together the segmented units and the tensile reinforcement form a block and thus is a gravity retaining structure.

seismic: referring to earth shaking due to earthquakes. Seismic type vibrations can also be caused by blasting, reciprocating machinery or construction activities.

shale: an earth material formed from clay deposits, which becomes hardened through pressure. Shale typically is layered or "fissile" and is a metamorphic type or rock.

shear: this is a force designation used in structural and geo-technical engineering to indicate forces that attempt to move materials against each other in a side to side direction.

Shelby Tube: a thin wall cylindrical sampling tube approximately 3 inches diameter with a sharpened cutting edge pressed into the earth by a drill rig to obtain nearly undisturbed samples of fine grain soils.

shoring: installation of temporary construction facilities used to keep an excavation or trench from caving in. It may also be used to temporarily support foundations or other parts of a structure while work is being performed.

shrinkage: the property of a soil material or other materials to reduce in volume because of the reduction of moisture content.

sieve: a device used in laboratory testing of soil or aggregates which has specified size of openings that permit only particles smaller than its size to pass through. A stack of sieves of different sizes will segregate a soil sample into its component sizes.

silt: a fine grain soil material ranging in size from that passing the #200 sieve down to clay sizes. Clay sizes are smaller than 0.002 mm.

skin friction: the resistance of a drilled pier or pile to penetration into the earth based on the side friction along the long side of the element. Different soil types have different amounts of skin friction available.

slabs-on-ground: a type of foundation construction which is commonly used for smaller buildings and consists of a reinforced concrete slab placed directly on the ground for support with or without stiffening beams or ribs.

slope: a ground surface, either natural or man made, that is not level. The term is also used to describe the amount of the change in surface elevation divided by horizontal distance. For example, a slope of 1v:20h would be a slope of a one-foot fall in 20 feet or 5 percent.

slump: when used in concrete technology this refers to the number of inches a stack of fresh concrete will settle after being placed in a tapering cone form after the form is removed. This is a measure of the workability or consistency of the concrete and is also used to control the amount of water that has been added to the concrete. If the slump is to little the concrete will not work readily in the forms and excessive honeycombing can result. If the slump is too much it may indicate that excessive water has been added and the strength and other properties of the concrete will suffer.

slurry: the mixture of water with earth materials or chemicals to produce a viscous liquid often used to stabilize drilled pier holes or test borings from caving.

sonic pulse procedure: this procedure is used in various aspects of construction to determine the integrity of typically a concrete unit. It also may be used on steel structural shapes. A sound or impact pulse is sent into the unit longitudinally and the sonic pulse device measures its speed of return thereby indicating if the unit is separated prior to the end of the unit being reached.

specification: a written and often a quite detailed description of a material or a process in a construction project used to supplement the drawings or plans.

split spoon sampler: a sampler used to obtain somewhat disturbed samples of the soil by pounding it into the ground. The number of blows to pound the sampler into the ground using a specified weight and drop can be used to evaluate the density or compaction of a sand or fine gravel deposit.

spring: A locality where underground water exits the surface and can be seen. Springs may be localized or over a wide spread area which are often called seepage areas. Springs may occur continuously, but may be seasonal or temporary after major rain events.

stacked rock gravity wall: a retaining structure which resists the lateral forces of the soil load by dead weight. A stacked rock gravity wall typically has a width of about $^1/_3$ of the height and relies on gravity and internal friction between the rock to provide the resistance. The rocks are usually nominal six to twelve inch irregular cut stones without mortar.

structural: this refers to the analysis and design procedures to provide sufficient strength to a constructed feature to resist the loads that will be imposed upon it. Such activities for major construction are usually performed by structural engineers.

subgrade: the plane that defines the separation of the original earth from fills, concrete base material, pavement layer or a drainage material in construction. The material below this plane is typically referred to as the subgrade soil.

subsidence: typically used to refer to a regional or local area which goes down as a result of natural processes often increased by withdrawal of groundwater or petroleum.

suction: the property of unsaturated soils and other materials to possess a negative energy which indicates the ability to draw moisture into the mass from areas of low suction to areas of high suction. It is typically measured in pF.

superstructure: that portion of a structure which is above the foundation elements.

thermal expansion or shrinkage: the reduction or increase in dimension of a material of construction such as concrete or steel caused by increase or reduction of temperature.

tilt: with reference to foundations this is the overall or planar movement of a foundation out of a level condition. Tilt is not the same as curvature.

topography: the physical attributes of the land, including its slopes, hills, and valleys. A topographic map is one that is frequently needed in site planning and design and contains an illustration of the land surface, typically by contour lines.

toppling: the falling of pieces of a cliff, typically along an ocean or lake front due to softening or erosion at the base. Toppling can be relatively minor consisting of boulders coming off the cliff or can be very major consisting of units of rock and soil weighing hundreds of tons.

triaxial compression test: laboratory soil strength tests in which confining pressure is placed all around the specimen in the testing machine before vertical load is applied. Similar to the unconfined compression test.

tremie: a hose or pipe used to transport concrete downward into forms or a drilled pier to the point of installation without having to free fall an excessive amount. A tremie may also be used to place concrete under water or slurry.

unconfined compressive strength: this is the strength of soil or a rock core when tested in a press. The results are recorded in pounds or tons per square foot at failure.

underpinning: this refers to elements that support a structure, sometimes temporarily or permanently during remediation or nearby construction. Construction people often refer to underpinning as the plastered closure between the floor of a building and the ground as underpinning as a term of art.

unsaturated soils: these are soils which do not have all their void spaces filled or nearly filled with water, leaving some voids filled with air.

USCS: Unified Soil Classification System. This system is used to classify the properties of soil of boulder to clay sizes by various methods of separating the percentages of sizes and the plasticity of the fine grain portion using Atterberg Limits. The USCS is a classification system based only on the raw material property of the soil and does not depend on its moisture condition or compaction.

waterproofing: the method of applying membranes or mastic on the exterior to typically a below grade wall to prevent water from entering a below grade area such as a basement. Damp proofing is a variant of this procedure without as much care for the details of the construction. Neither waterproofing nor damp proofing can totally prevent a head of groundwater from penetrating into an interior space, and the water head should be reduced by subsurface drainage.

weighted average: this is the average of a collection of data which takes into account the extent of a particular value compared to the total extent. Specifically used with regard to soil profiles and seismic site evaluation in that the thicker a layer is with a particular property the more it adds to the weighted average.

welded wire fabric: a form of reinforcing steel in which wires typically smaller than the smallest reinforcing bars are welded together to form a grid pattern for use in concrete construction. Welded wire fabric is designated by the spacing between the wires and the size of the wires, sometimes with different spacing and sizes in two directions. The material can be obtained in rolls or in sheets.

well graded: in soils engineering this refers to the assemblage of soil particles which have fairly constant percentages of each size within the range present as opposed to poorly graded or gap graded, which has sizes missing or in excessive amounts.

WRI: Wire Reinforcing Institute

TEST QUESTION ANSWERS

CHAPTER ONE
INTRODUCTION

		Book Reference	2012 Code Reference
1.	a, c, d	Section 1.4	
2.	a, b, d	Section 1.4	
3.	a	Section 1.2	
4.	b, c, d	Section 1.3	
5.	b, c	Section 1.3	
6.	b	Section 1.3	
7.	b	Section 1.4	
8.	a, c	Section 1.4	
9.	b	Section 1.1	IBC Cp. 35, IBC 102.4
10.	b, c	Section 1.4	
11.	b	Section 1.4	IBC Preface page ix

CHAPTER TWO
THE PURPOSE OF FOUNDATIONS AND A FOUNDATION'S RELATIONSHIP TO SOIL

		Book Reference	2012 Code Reference
1.	b, d	Section 2.1.2 Figs. 2.3, 2.4	
2.	a, d	Section 2.3	
3.	b	Section 2.3	
4.	b	Section 2.1	
5.	b	Section 2.1.2	
6.	c	Section 2.1.2	
7.	b, c	Section 2.1.3	
8.	a	Section 2.2	
9.	a	Section 2.2	
10.	b	Section 2.2	
11.	b	Section 2.3 Table 2.1	IBC Table 1607.1

CHAPTER THREE
ROCK

		Book Reference	2012 Code Reference
1.	b	Section 3.1	
2.	c	Section 3.2	
3.	c	Section 3.2	
4.	a, c	Section 3.3	
5.	c	Section 3.3	
6.	b	Section 3.4	IBC Table 1806.2
7.	a	Section 3.2 Fig. 3.1	
8.	b	Section 3.2	
9.	c	Section 3.4	
10.	c	Section 3.4 Table 3.1	IBC Table 1806.2

CHAPTER FOUR
SOIL

		Book Reference	2012 Code Reference
1.	b	Section 4.4	
2.	b	Section 4.7	
3.	c	Section 4.1	
4.	b	Section 4.1	
5.	c	Section 4.2 Fig. 4.4	IBC 1803.5.1 IRC Table R405.1
6.	b	Section 4.2	
7.	c	Section 4.2	
8.	b	Section 4.2 Table 4.1	IRC Table R405.1
9.	c	Section 4.4	
10.	c	Section 4.1	
11.	d	Section 4.8	IBC 1809.5
12.	c	Section 4.8	IRC Table R301.2 (1), footnote "b"

CHAPTER FIVE
GROUNDWATER AND DRAINAGE

		Book Reference	2012 Code Reference
1.	c, d, e	Section 5.3.2	
2.	c	Section 5.3.1	
3.	c	Section 5.3.3	IBC 1805, IRC R403.1.7 and R405
4.	b	Section 5.3.1	
5.	a, b	Section 5.1.1	
6.	c	Section 5.1.1	
7.	c	Section 5.3.2	IBC 1804.3, IRC R401.3

CHAPTER SIX
SITE INVESTIGATIONS

		Book Reference	2012 Code Reference
1.	a, c	Section 6.3	
2.	b, c	Section 6.1	IBC 1803, IRC R401.4
3.	a, c, d	Section 6.2	
4.	b	Section 6.2	
5.	a, c	Section 6.3	
6.	c	Section 6.3	
7.	b	Section 6.3	
8.	b	Section 6.4	
9.	b	Fig. 6.9	
10.	c	Section 6.6	
11.	b	Section 6.9	

CHAPTER SEVEN
TESTING SOIL AND ROCK

	Book Reference	2012 Code Reference
1. b	Section 7.3.2.1	
2. a, d, e	Section 7.3.2.2	
3. a, c	Section 7.3.1	
4. b	Section 7.3.1.2	
5. b	Section 7.3.2.2 Fig. 7.9	
6. b	Section 7.3.3.3	
7. b	Section 7.4.5	
8. c	Section 7.2.2	
9. d	Section 7.3.1.2	
10. c	Fig. 7.6	
11. d	Section 7.3.1.4	
12. d	Table 7.1	
13. c	Table 7.2	
14. b	Section 7.3.2.3	IBC 1808.6.2
15. d	Table 7.3	

CHAPTER EIGHT
ANALYSIS OF SITE INFORMATION AND CONSTRUCTION DOCUMENTS

	Book Reference	2012 Code Reference
1. c, d, f	Section 8.1	IBC 1803.6
2. b	Section 8.4	
3. c	Section 8.5	
4. a	Section 8.6	
5. c, d	Section 8.2	
6. b	Section 8.2	
7. c	Section 8.3	
8. b	Fig. 8.1	
9. a, c	Section 8.8	
10. b	Section 8.11	
11. c	Section 8.12	IBC 1803.2, IRC R401.4
12. b	Section 8.13.1	
13. b	Section 8.13.3	
14. a	Section 8.13.6	

CHAPTER NINE
EXCAVATION AND GRADING

		Book Reference	2012 Code Reference
1.	b	Section 9.3	
2.	b	Section 9.6	
3.	b	Section 9.10	
4.	a, b	Section 9.1	
5.	b, d, e	Section 9.1	IRC R403.1.7, IBC Appendix J, IBC 1804 & 1808.7
6.	a, b	Section 9.1	IBC 1804.4
7.	c	Section 9.2	
8.	b	Section 9.2	
9.	a	Section 9.2	
10.	b	Section 9.2	
11.	c	Section 9.2 & 9.5.6	
12.	c	Section 9.2	
13.	b	Section 9.5.6	
14.	c	Section 9.6	
15.	b	Section 9.5.5	
16.	b, c, d	Section 9.5.7.1	
17.	b	Section 9.5.7.1	
18.	a	Section 9.5.1	

CHAPTER TEN
SOIL AND SEISMICS

		Book Reference	2012 Code Reference
1.	b	Section 10.1	
2.	a	Table 10.1	2006 IBC Table 1613.5.2 & IBC 1613.2 & ASCE 7 Table 20.3.1
3.	b	Section 10.2 (1)	2006 IBC Table 1613.5.2
4.	b	Section 10.1	2012 IRC 1803.3, 1803.5.11, 1803.5.12
5.	c	Section 10.2 (2)	2006 IBC Table 1613.5.2 & ASCE 7 Table 20.3.1
6.	e	Table 10.1	2006 IBC Table 1613.5.2 & ASCE 7 Table 20.3.1
7.	b	Section 10.2 (7)	2012 IBC 1803.5.11 & 1803.5.12

CHAPTER ELEVEN
SPREAD OR STRIP FOOTINGS

		Book Reference	2012 Code Reference
1.	a	Section 11.4	
2.	c	Section 11.1	IRC R403, IBC 1808 & 1809
3.	c	Section 11.1	
4.	b, c, e	Section 11.1	IRC R403, IBC 1808 & 1809
5.	c	Section 11.1	
6.	c	Section 11.4	IRC R403.1.5, IBC 1809.3
7.	a, c	Section 11.2	
8.	b	Section 11.3	
9.	d, e, f	Section 11.3	IBC 1809.13
10.	d	Section 11.4 Fig. 11.2	IRC R403.1.7, IBC 1808.7

CHAPTER TWELVE
PIER FOUNDATIONS

		Book Reference	2012 Code Reference
1.	a, b	Section 12.1	IBC 1810
2.	c	Section 12.3	
3.	b	Section 12.3.2	
4.	b	Section 12.5.3	
5.	c	Section 12.5.4	
6.	b	Section 12.3	
7.	b	Section 12.3	
8.	c	Section 12.3	
9.	d	Section 12.3.2	
10.	c	Section 12.4	
11.	c	Section 12.5.1	

CHAPTER THIRTEEN
PILE FOUNDATIONS

	Book Reference	2012 Code Reference
1. b, c, e	Section 13.2	IBC 1810
2. b, c, e	Section 13.4	
3. c	Section 13.1	
4. b	Section 13.1	
5. a	Section 13.3.1	
6. c	Section 13.3.1	
7. b	Section 13.3.2 Fig. 13.1	IBC 1810.3.3.1.2
8. b, c	Section 13.4	
9. d	Section 13.3.2	
10. b	Section 13.3.2	IBC 1810.3.3.1.2

CHAPTER FOURTEEN
RAFT OR MAT FOUNDATIONS

	Book Reference	2012 Code Reference
1. c	Section 14.3.3 Fig. 14.6	
2. b	Section 14.5	
3. b	Fig. 14.4	
4. b	Section 14.3.2	
5. b	Section 14.4	
6. c	Section 14.5	
7. b	Section 14.7	
8. a, b	Section 14.7	
9. b	Section 14.3.1	
10. c	Section 14.3.3	IBC 1808.6
11. b	Section 14.4	

CHAPTER FIFTEEN
SOIL-SUPPORTED SLAB FLOOR WITH STRUCTURAL FOOTINGS

		Book Reference	2012 Code Reference
1.	b	Section 15.1	
2.	b	Section 15.3	
3.	b	Section 15.4	
4.	b	Section 15.3	
5.	c	Section 15.3	
6.	c	Section 15.3	
7.	c	Section 15.4	
8.	c	Section 15.4	

CHAPTER SIXTEEN
SITE-STABILIZATION TECHNIQUES

		Book Reference	2012 Code Reference
1.	a, c, e	Section 16.2 & 16.2.2 & Section 16.3	IBC 1808.6.3 & 1808.6.4
2.	a, b	Section 16.1	
3.	b	Section 16.3	
4.	c, d	Section 16.2.2 & 16.2.1	
5.	c	Section 16.2.2	
6.	a	Section 16.3	
7.	c	Section 16.3	
8.	c	Section 16.3	
9.	b, d	Section 16.4 & 16.5	
10.	b	Section 16.4	

CHAPTER SEVENTEEN
RETAINING STRUCTURES

		Book Reference	2012 Code Reference
1.	a	Table 17.1	IBC Table 1610.1
2.	b	Section 17.4.2 & Fig. 17.8	IBC Table 1806.2
3.	c	Section 17.5	
4.	b	Section 17.5	
5.	b	Section 17.1	IRC R404, IBC 1610 & 1806 & 1807.2
6.	b, d	Section 17.2	
7.	c	Section 17.2	
8.	b	Section 17.2	
9.	c	Fig. 17.5	
10.	c	Section 17.4.1 & Fig. 17.7	
11.	b	Section 17.4.1 & Table 17.1	IBC Table 1610.1

CHAPTER EIGHTEEN
SLOPE STABILITY

		Book Reference	2012 Code Reference
1.	a, b	Section 18.3.1	
2.	b	Section 18.3.2	
3.	b, c, e	Section 18.3.1	
4.	a, c, f, g	Section 18.3.1 & 18.3.2 & Section 18.3.3 & 18.3.4	
5.	b	Section 18.3.3	
6.	c	Section 18.3.4	
7.	c	Section 18.3.3	
8.	c, e	Section 18.3.4	

CHAPTER NINETEEN
RECAP OF SITE AND FOUNDATION "RED LIGHTS"

		Book Reference	2012 Code Reference
1.	b	Section 19.1.4	
2.	b	Section 19.1.5	IBC 1613
3.	a	Section 19.1.6	
4.	b	Section 19.1.7	
5.	c	Section 19.1.9	
6.	a, b	Section 19.2	
7.	d	Section 19.1.1	
8.	a, b	Section 19.1.3	
9.	c	Section 19.1.8	

CHAPTER TWENTY
CONSTRUCTION INSPECTION

		Book Reference	2012 Code Reference
1.	b	Section 20.1, 20.2	IBC Table 1705.3, IBC Table 1705.6, IBC Table 1705.7, IBC Table 1705.8
2.	b	Section 20.6	
3.	b	Section 20.3.1, 20.3.8	
4.	b	Section 20.3.1	
5.	a	Section 20.3.7	
6.	b	Section 20.3.5	
7.	c	Section 20.3.6.2 Fig. 20.2	
8.	b	Glossary, Fig. 20.6, Fig. 20.7	
9.	d	Section 20.3.2	
10.	a, c, d	Section 20.3.3	IBC Table 1705.7
11.	c	Section 20.3.6.2	
12.	b, d	Section 20.3.6.2	
13.	d	Section 20.3.8	IRC R401.3, IBC 1804.3

CHAPTER TWENTY-ONE
FOUNDATION NON-PERFORMANCE

		Book Reference	2012 Code Reference
1.	d	Section 21.2	
2.	a	Section 21.3	
3.	b	Section 21.3	
4.	c	Section 21.4	
5.	b, d, e	Section 21.1	
6.	a, b, c	Section 21.1	

Appendix
EXAMPLE OF SOILS REPORT

GEOTECHNICAL INVESTIGATION
FOUNDATION RECOMMENDATIONS

██████████ Elementary School
██████ County, Texas

Report For:

██████████ ISD
Texas 78640

November 2006

Engineer's Job # ████████

MLA LABS, INC.
Geotechnical Engineering and
Construction Materials Testing
"put us to the test"

Geotechnical Engineer

Vice President   11|15|06

 P.E.
Senior Consultant

Elementary School
Engineer's Job No.

TABLE OF CONTENTS

MLA LABS, INC.   2804 LONGHORN BLVD.   AUSTIN, TEXAS 78758   512/873-8899   FAX 512/835-5114

GEOTECHNICAL INVESTIGATION
Foundation Recommendations

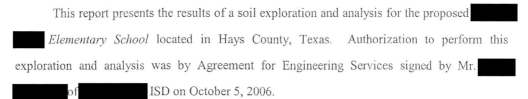
Elementary School
Texas

BACKGROUND

This report presents the results of a soil exploration and analysis for the proposed ▮ ▮ *Elementary School* located in Hays County, Texas. Authorization to perform this exploration and analysis was by Agreement for Engineering Services signed by Mr. ▮ ▮ of ▮ ISD on October 5, 2006.

The purposes of this investigation were to determine the soil profile and the engineering characteristics of the foundation soil, and to provide criteria for use by the design engineers in preparing the foundation and pavement designs for the proposed project. The scope included a review of geologic literature, a reconnaissance of the immediate site, a subsurface exploration, field and laboratory testing, and an engineering analysis and evaluation of the foundation materials.

The exploration and analysis of the subsurface conditions reported herein is considered sufficient in detail and scope to form a reasonable basis for foundation and pavement designs. The recommendations submitted are based on the available soil information and the assumed preliminary design for the proposed project. Any revision in the plans for the proposed structure from those stated in this report should be brought to the attention of the Geotechnical Engineer so that he may determine if changes in the foundation recommendations are required. Site work and foundation construction should be monitored by MLA Labs, Inc. to verify that these recommendations are implemented, and so that deviations from expected conditions can be properly evaluated.

This report has been prepared for the exclusive use of the client and their design professionals for specific application to the proposed project in accordance with generally

---

MLA LABS, INC.   2804 LONGHORN BLVD.   AUSTIN, TEXAS 78758   512/873-8899   FAX 512/835-5114

██████ Elementary School
Engineer's Job No. ██████

accepted soils and foundation engineering practice. This report is not intended to be used as a specification or construction contract document, but as a guide and information source to those qualified professionals who prepare such documents.

ARCHITECTURAL AND STRUCTURAL ASSUMPTIONS

The proposed structure is a single story school building with associated parking areas and access ways. Details concerning the construction techniques and materials were not available at the time of this report. Estimates of structural loads were also unavailable. As finalized plans become available they should be shared with the Geotechnical Engineer so they may ascertain whether any modifications to the recommendations presented herein are necessary.

FIELD AND LABORATORY INVESTIGATION

Sixteen borings were drilled to various depths spaced at locations as shown on the enclosed Logs of Boring and Plan of Borings using a B-53 drilling rig. Water was not introduced into the borings. The field investigation included completing the soil borings, performing field tests, and recovering samples. Pocket penetrometer tests were performed on specimens during sampling. Representative soil samples were selected for laboratory index tests including Atterberg Limits and moisture content tests. The results of these tests and stratigraphy are presented on the Logs of Boring found in *Appendix A*. A key to the Soil Classification and symbols is located behind the last Log of Boring. See *Appendix B* for details of field and laboratory procedures, as applicable.

SITE TOPOGRAPHY, DRAINAGE AND VEGETATION

The topography of this site is slight with existing grades of approximately 1 percent. Regionally, this site drains to the south. Prior to development, the vegetation at this site included primarily grasses but included a moderately treed area.

-2-

█████ Elementary School
Engineer's Job No. █████

SUBSURFACE CONDITIONS AND LOCAL GEOLOGY

Site Profiles

The soil profile identified in the borings generally consists of an upper layer of dark brown high plasticity clay (CH) underlain by pale brown to light yellowish brown low plasticity clay (CL) that varies to a clayey or silty sand (SC to SM). This low plasticity clay or sand stratum contains variable amounts of gravel. Moist to wet ground conditions were noted in these gravel lenses. In many borings, this lower sand section becomes cemented and is thought to vary to sandstone and conglomerate. The upper portion of this soil profile has the potential for volume change with varying soil moisture contents.

Geology

Geologic maps indicate an outcropping of Terrace Deposits, *Qt*, at this site [1,2]. The terrace deposits at this site are part of a vast network of flood plains of ancient rivers that covered much of central Texas during the Quaternary Period. The composition of these terrace deposits was dependent on the depositional energy of the flood plain and the composition of the material eroded by the rivers. Locally, terrace deposits from the Quaternary Period are quite variable horizontally due to the nature of fluvial systems.

The older portion of these terrace deposits generally consists of coarse sand with thick layers of coarse gravel. Often, very little clay is present in the lower (oldest) terrace deposits. As the geologic record progresses to the present, the sand becomes finer and gravel layers become less frequent and thinner and the gravel sizes are smaller. Larger proportions of clay are typically realized in the upper portions of terrace deposits. These terrace deposits typically terminate in the Pleistocene Epoch where they consist of fine sand and clay. More recent alluvial material is not often mapped separately from the older terrace deposits because the boundary is not sufficiently distinct.

-3-

████████ Elementary School
Engineer's Job No. ████████

Calcareous zones occur occasionally in terrace deposits and are thought to be caused by the precipitation of salts in ground water from extreme drying periods. The extreme drying periods resulted in the sediments of that time being over-consolidated. Terrace deposits generally have moderate bearing capacity and a variable shrink and swell potential. Due to their nature, terrace deposits have the ability to store and transmit large quantities of ground water.

Faults

Published geology maps do not indicate the presence of a fault on the project site and faulted conditions were not noted in the borings.

Ground water

Free ground water was not encountered in the borings during this investigation. However, moist ground conditions were noted at depth. This formation possesses the potential for producing varying quantities of ground water. It should be noted that groundwater may be encountered at other times, depths and locations depending on the antecedent rainfall and changes in land use.

-4-

███████ Elementary School
Engineer's Job No. ███████

CONCLUSIONS

1. Excavation and site work considerations:

   a. Excavation may be performed in the surface soils using ordinary power equipment. Pier holes may require equipment designed for medium to hard sandstone.

   b. All excavations should be braced and shored according to applicable law and building code. Consultation on excavations can be provided by the geotechnical engineer upon request. If shoring is required on this project, specific design recommendations can be developed upon analysis of the application.

   c. Ground water is possible in shallow and deep excavations depending on antecedent rainfall. During periods of high rainfall, perched ground water may cause the soils to become soft and difficult to compact.

2. Settlement potential:

   a. Settlement potential of the natural soils on this site for light, one to three story structures may be categorized as negligible.

   b. Heavy structures or structures more than three stories in height will require analysis beyond the scope of this report.

3. Expansive soil potential:

   The soils at this project site exhibited plasticity indices ranging from 10 to 56. Point estimates of the potential vertical rise, PVR, of the in-situ soil profile were found to range from 2-¼ inches to 3 inches [3]. Thus, the potential for disruptive foundation movements due to swelling soils may be categorized as high. Other magnitudes of PVR may be estimated by other methods and at other locations with varying results. However, the TxDOT Method is widely used and should be considered an index property of the site. PVR is considered in the final foundation recommendations.

-5-

████████ Elementary School
Engineer's Job No ████████

4. Foundation Type:

    The foundation type recommended for this project is a pier foundation system with beams supported clear of grade. Options are presented for the slab portion of the structure to be suspended clear of grade or founded on compacted select fill. Please see *Recommendations – Foundation*.

5. Faults:

    Published geology maps do not indicate the presence of a fault on the project site and faulted conditions were not noted in the borings.

6. Slab Moisture:

    The recommendations in this report are not intended to address the effects of moisture migration through slabs. The design team should address moisture retardant schemes and the requirements of this project.

7. Past Use of Site:

    There was no evidence in the samples obtained for this study that indicated the past use of this site as a municipal landfill. See the section *Limitations of Report*.

8. This site is a Class D site with a "Stiff soil profile" as per Table 1615.1.1 of the International Building Code 2003 [4].

-6-

██████Elementary School
Engineer's Job No██████

RECOMMENDATIONS - FOUNDATION

The foundation type recommended for this project is a pier foundation system with beams supported clear of grade. Options are presented for the slab portion of the structure to be suspended clear of grade or founded on compacted select fill.

Pier Design Recommendations

1. The foundation elements should be supported by drilled straight-shaft concrete piers established a minimum of 20 feet below the current ground surface. Due to the variable subsurface conditions encountered we have developed the following table of pier capacities. If additional capacity is needed, please contact the geotechnical engineer.

Bearing Capacity of Individual Pier*

Pier Diameter, Inches	Pier Capacity, Kips
18	35
24	47
30	58
36	70
42	82
48	94

*Pier founded 20 feet below the existing ground surface.

2. Please be aware that the conditions encountered during pier drilling will be highly variable, with some piers drilled easily and other piers drilled with more difficulty through layers of sandstone.

3. Uplift forces acting along the shaft of the drilled pier in clay can be conservatively estimated at 4,000 psf in the upper CH clay portion of the soil profile. This does not represent the net upward force, but only the component due to the action of the soils. The

-7-

███████ Elementary School
Engineer's Job No. ████████

weight of the pier and the skin friction developed in the 10 foot to 20 foot depth below the ground surface are expected to be sufficient for overcoming uplift forces at this site.

4. The minimum area of vertical steel reinforcement should be as per structural analysis, considering the soil uplift and building loads.

5. Any piers that may be in tension should be checked for pullout resistance.

6. Casing may be required to install drilled piers in the event that groundwater is encountered or in the case where sloughing soils are encountered. Both moist soils and granular soils that tend to collapse un-cased pier holes were encountered during drilling at this site. In addition, the potential for ground water is strongly dependant on the time of year and antecedent rainfall prior to and during construction. Contract documents should provide for an add/deduct line item for the provision of temporary steel casing.

Slab Option 1 – Slab Suspended Clear of Ground Surface

1. The grade beams should be <u>structurally</u> connected to the drilled piers to provide a <u>fixed</u> connection.

2. Grade beams and slab should be separated from the soil by a minimum 12-inch void formed underneath the foundation elements. The void space should be graded to provide crawl-way access (if desired) and positive drainage of any water that collects under the foundation. Wood or steel forms are best used to create this 12-inch void beneath the grade beams and slab because of the tendency of cardboard forms to collapse during construction. However, cardboard forms may be used. Positive post-construction verification of the void space should be performed.

3. The floor may be designed of steel, concrete, or wood or combination of the three. The floor should span clear of the soil surface by a minimum distance of 12 inches and adequately vented as per applicable building code. Wood may require an increased clearance from the soil to reduce the risk of rotting and termite damage.

4. All plumbing stacks should be provided with a swing joint.

-8-

MLA LABS, INC.   2804 LONGHORN BLVD.   AUSTIN, TEXAS 78758   512/873-8899   FAX 512/835-5114

Elementary School
Engineer's Job No

5. The underslab void space must be well ventilated, as per applicable building codes. Passive ventilation systems are preferred over powered systems because of the reduced maintenance required. Forced ventilation systems should be properly maintained. A qualified expert should review the adequacy of underslab ventilation.

Slab Option 2 – Slab Founded on Compacted Select Fill

1. Prepare the subgrade and provide a building pad as per the following:

    a. Strip and remove all surface dark brown CH clay from beneath the building pad. This should consist of a minimum 36-inch deep excavation. Due to the variability in subsurface conditions encountered at this site completed, additional removal of CH clay may be required after the 36-inch excavation is complete. Contact the geotechnical engineer to observe the completed excavation.

    b. Replace the removed soil and make up required grades with select fill as per the enclosed *Select Fill Recommendations*. A PVR of 1.0 inch or less is the industry recognized standard for this type of foundation and should be achieved using the remove and replace procedure outlined herein.

    c. The removal of native material and placement of fill should be done to a neat line coinciding with the outside edge of the foundation to reduce surface water infiltration below the foundation. The backfill for grading around the exterior of the completed foundation should consist of a minimum of 12 inches of the high plasticity clay removed during the excavation process.

2. A moisture retardant layer of sealed, overlapping plastic sheeting should be provided between the subgrade and all slab areas and beam excavations to retard the transmission of moisture upward through the slab. ASTM 1745 should be used as a guideline [6]. The Client and the Contractor should address moisture retardant schemes and the requirements of this project.

-9-

MLA LABS, INC.   2804 LONGHORN BLVD.   AUSTIN, TEXAS 78758   512/873-8899   FAX 512/835-5114

███████ Elementary School
Engineer's Job No. ███████

3. Grade beams should be separated from the soil by a minimum 12-inch void formed underneath the foundation elements. Wood or steel forms should be used to create this 12-inch void beneath the grade beams. Cardboard void box forms are not recommended due to their tendency for collapse. The void space should be graded to provide positive drainage of any water that collects under the foundation. The void space should be verified after the completion of construction of the foundation.

4. The grade beams should be <u>structurally</u> connected to the drilled piers to provide a <u>fixed</u> connection.

5. The slab thickness for the building should be a minimum of 5 inches or as determined by structural analysis. Reinforcing and joint spacing for crack control shall be as per structural analysis.

MLA LABS, INC.   2804 LONGHORN BLVD.   AUSTIN, TEXAS 78758   512/873-8899   FAX 512/835-5114

██████ Elementary School
Engineer's Job No ████████

Miscellaneous Recommendations

1. Concrete should develop a minimum 28-day strength of 3,000 psi.

2. Flatwork and sidewalks at this site not structurally suspended on drilled piers along with the foundation slab will be susceptible to movements of the underlying soils. This movement may be reduced—but not eliminated—by providing 18 inches of compacted crushed stone beneath the flatwork. The crushed stone should be placed in accordance with the enclosed *Select Fill Recommendation*.

3. Irrigated landscaping should be maintained away from the building perimeter.

4. Drainage should be maintained away from the foundation, both during and after construction. Water should not be allowed to pond near the foundation. The following items should provide for positive drainage of water away from the foundation: sidewalks and other concrete flatwork, parking areas, driveways and other surface drainage features, landscaping (including irrigated landscaping), gutters, and air conditioner condensation overflow drains.

5. Upon completion of the plans and prior to construction, the Geotechnical Engineer should be given an opportunity to review the plans in order to assure that these recommendations have been properly included into the plans. During construction, it is recommended that the Geotechnical Engineer be given the opportunity to inspect the construction of the foundation to verify that these recommendations are followed.

-11-

_____ Elementary School
Engineer's Job No. _____

RECOMMENDATIONS - PARKING AND DRIVING LANE PAVEMENTS

No truck traffic loads or frequencies were available at the time this report was written. Therefore, pavement thickness sections are based on primarily passenger cars and light trucks with an average of one heavy-duty truck per day.

A.   Subgrade and Foundation Soil Preparation

     1.   Strip and remove from construction area all topsoil, organics and vegetation to a minimum depth of 6 inches. Any fill of unknown consistency should be removed and replaced in accordance with the enclosed *On-Site Fill Recommendations*, unless otherwise specified.

     2.   Proof-roll the subgrade in accordance with TxDOT Item 216 to reveal soft spots, remediating these soft spots with compact fill in accordance *On-Site Fill Recommendations.*

B.   Base Course

     1.   Base material shall be Type A, Grade 2 or better, according to the Texas Department of Transportation Specification Item 247.

     2.   Thickness of the base course should be in accordance with Table 1.

     3.   Base course compaction should be 100 percent of TxDOT TEX-113-E. Density control by means of field density determinations shall be exercised. The base course should be within 3 percent of optimum moisture at time of compaction.

     4.   Proof-roll the base course in accordance with TxDOT Item 216.

     5.   After compaction, testing and curing of the base material, the surface should be primed using an Asphalt Emulsion prime coat meeting TxDOT Specification Item 310.

     6.   A full thickness of the base course should be extended 3 feet beyond the back of curb line for high plasticity (CH) clay subgrades.

-12-

C. Flexible Pavement - Hot Mixed Asphalt Concrete

Surfacing shall consist of hot mix asphaltic concrete meeting the requirements of TxDOT Item 340, Type D mixture. Thickness should be in accordance with Table 1. The HMAC should be compacted to a minimum of 91 to 96 percent of the maximum theoretical density with all rolling completed before the HMAC temperature drops below 175° F.

D. Concrete paving shall consist of thickness as given in Table 1. As a minimum, rigid (concrete) paving should be used for dumpster pads and around dumpster loading areas as per Heavy Duty Truck Lanes (Average Daily Truck Traffic = 1). The concrete should develop a minimum 28-day flexural strength of 500 psi with 4 to 6 percent entrained air. The flexural strength of concrete may be approximated using the following formula taken from ACI 330R:

$$M_r = 2.3 f_c^{(2/3)}$$

where M_r = flexural strength of concrete in psi

f_c = compressive strength of concrete in psi

Minimum reinforcing should be No. 3 bars at 18" on center each way, centered in the slab or as determined by the following ACI "Drag Formula "[5].

$$A = \frac{L \cdot C_f \cdot w \cdot h}{24 \cdot f_s}$$

where A = area of distributed steel reinforcement required per foot of slab (in.^{2})

L = distance between joints (feet)

C_f = coefficient of subgrade resistance to slab movement (a value of 1.5 is most commonly used in design)

w = weight of concrete (145 lb/ft^3)

h = slab thickness (in.)

f_s = allowable tensile stress in distributed steel reinforcement, psi (a value of 2/3 yield strength is commonly used)

Contraction, control, and expansion joint details should be determined in accordance with guidelines published by the American Concrete Institute[5], the Portland Cement Association[6] or accepted local practice that has been proven to work satisfactorily in similar circumstances. Contraction joint spacing should not exceed 20 feet on center

-13-

███████ Elementary School
Engineer's Job No ███████████

without engineering consultation. Full depth, full width isolation joints with bituminous fiber or preformed joint filler should be installed at all rigid structure interfaces, such as light pole bases, planters and buildings or older sections of pavement.

E. General Conditions

1. Should at any stage in the construction of the pavement a non-stable or weaving condition of the subgrade or base course be noted under the wheel loads of construction equipment, such areas should be delineated and the Geotechnical Engineer consulted for remediation before completing the pavement section.

2. Seepage areas or unusual foundation soil conditions should be similarly brought to the Geotechnical Engineer's attention before proceeding with pavement completion.

3. Landscaped islands should be backfilled with low plasticity clays to reduce water intrusion into the subsurface pavement structures. Curbs should be provided with weep holes in landscaped areas to reduce the build up of hydrostatic pressure and to reduce the intrusion of water into the subsurface materials.

4. Trenches beneath pavements should be backfilled with borrow or suitable material excavated from the trench and free of stone or rock over 8 inches in diameter. The backfill should be compacted to 95 percent of the maximum dry density when determined by TxDOT test method Tex-114-E. The moisture content should be within 2 percent of the optimum moisture content at the time of compaction.

5. If ground water or seepage is encountered at the time of construction, French drains may be required to drain or intercept the flow of water from the subsurface pavement materials. These drains should be sloped a minimum of 0.5 percent to provide positive drainage to daylight.

-14-

██████ Elementary School
Engineer's Job No. ██████

TABLE 1 : Recommended Pavement Section Thickness, Inches

Expected Traffic	Average Daily Truck Traffic	Flexible Pavement		Rigid Pavement	
		HMAC	CLB	JRPCC	CLB
Passenger Vehicles	0	1.5	8	5	-
Heavy Duty Trucks	1	2.0	10	6	-
Heavy Duty Trucks	10	-	-	6	6

Notes:

- Abbreviations: HMAC -Hot Mixed Asphalt Concrete, CLB - Crushed Limestone Base, JRPCC - Jointed, Reinforced Portland Cement Concrete
- Average Daily Truck Traffic excludes pickup and panel trucks.
- Where greater than 2 feet of high plasticity clay will form the subgrade, the thickness of CLB should be increased by 50% for flexible pavements only. This increase would be a total CLB thickness of 12 inches for passenger vehicles and 15 inches for heavy duty trucks.
- Inadequate drainage of the pavement system will accelerate pavement distress and result in increased maintenance costs. Adequate drainage should be provided for the pavement system. Adequate drainage consists of a curb and gutter or a shoulder and bar ditch system. The final pavement cross section and drainage should be reviewed by the geotechnical engineer.
- These pavement thickness designs are intended to transfer the load from the anticipated traffic conditions. Deep seated soil swelling or settlement of fill materials may cause long wave surface roughness. The recommendations above are intended to reduce maintenance costs and increase the serviceable lifespan of the pavement system.

-15-

███████ Elementary School
Engineer's Job No. ██████████

ON-SITE FILL RECOMMENDATIONS

A. Selection of on-site fill material shall be guided by the following criteria:

    1. The material shall not contain any rocks having a maximum dimension greater than six inches.

    2. The material shall have at least 50% passing the #4 sieve. The soil material passing the #40 sieve shall have a plasticity index of 20.

    3. The material shall be reasonably free of roots, trash, concrete rubble and other organic material.

B. Compaction shall be to 95 percent of maximum laboratory density determined in accordance with ASTM D 698. The material shall be within three percent of optimum moisture content during compaction.

C. Placement shall be in lifts not exceeding eight inches after compaction. Each compacted lift should be inspected and/or tested for density compliance by the Geotechnical Engineer prior to placing the next lift. The fill area should extend at least 24 inches (36 inches on fills over six feet in height) beyond the back of curb or foundation line before sloping downward on not more than 1 on 2 slope to natural soil. Backslopes shall be well-compacted. Maximum fill heights should not exceed ten feet without engineering consultation.

D. Testing and Certification of the on-site fill material shall be performed by the Geotechnical Engineer. A 110 lb. sample of the proposed material shall be submitted to the Geotechnical Engineer for approval and determination of a moisture-density relationship in advance of the fill and compaction operations in order to permit inspection and testing as the fill is placed. Fill placement will be inspected and tested for uniformity, acceptable material and field densities at the rate of one per density per 5,000 square feet per lift (a minimum of three per lift per pad).

E. Deviations from the above recommendations may be permitted upon approval from the Geotechnical Engineer.

MLA LABS, INC.   2804 LONGHORN BLVD.   AUSTIN, TEXAS 78758   512/873-8899   FAX 512/835-5114

Elementary School

Engineer's Job No.

SELECT FILL RECOMMENDATIONS

1. **GENERAL**: Select fill, if called for on the plans, shall be placed over prepared compacted foundation soil to the dimensions shown on the plans.

2. **MATERIAL**: Select fill material shall be composed of hard durable particles of gravel or crushed stone and shall meet the following criteria:

   A. Gradation shall be as follows:

Sieve Size	Percent Finer by Weight
1-3/4"	100
1-1/2"	85 - 100
3/4"	45 - 75
No. 4	25 - 70
No. 40	10 - 40

   B. Material passing the No. 40 sieve shall meet the following:

Percent Passing No. 40	Max. PI	Min. PI
25 - 40	15	3
10 - 25	20	4

   C. Maximum liquid limit of the minus No. 40 material shall be 35.

   D. No organic matter is permitted.

3. **PLACEMENT AND COMPACTION**: Compaction should be to 95 percent of maximum laboratory density determined in accordance with American Society of Testing Materials, Method ASTM D 698. Material should be within three percentage points of optimum moisture at time of compaction.

   Placement should be in lifts not exceeding six inches after compaction. Each compacted lift should be inspected and tested for density compliance prior to placing the next lift.

   After completion, not less than plan thickness of select, compacted fill as herein recommended shall exist beneath any portion of the foundation, even if additional excavation of existing ground is required to meet this requirement.

4. **INSPECTION, TESTING AND CONTROL**: A 110 lb. sample of proposed fill material should be submitted to the Engineer for approval and for determination of Moisture-Density Relationship, at least seven days in advance of placement. Fill placement operations will be inspected and tested for uniformity, acceptable material and field densities, at the Engineer's option. Testing and inspection will be at the Owner's expense, or paid by allowance.

-17-

▇▇▇▇ Elementary School
Engineer's Job No. ▇▇▇▇▇▇

QUALITY ASSURANCE CONSIDERATIONS

Type of Work	Item	Sample Frequency	Sample Size	Minimum Testing
General Earthwork and Fill Material	Soil	1 per Soil Type	110 lbs.	◆ Sieve ◆ P.I. ◆ Moisture Density Relationship
	Compaction	1 per 5000 ft² per lift (min. of 3 per lift)		Field Density Test
Select Under-slab Fill	Select Fill Material	1 per type per 1000 cu. yds. Min. one per job	110 lbs.	◆ Sieve ◆ P.I. ◆ Moisture Density Relationship
	Compaction	1 per 2000 ft² per lift (min. of 3 per lift)		Field Density Test
Concrete or HMAC	Mix Design	1 per concrete class		◆ Review & approval with confirmatory cylinders ◆ Plant & materials approval, testing, if questionable
Concrete or HMAC	Aggregates (coarse & fine)	1 per 500 cu. yd. Min. 1 per job	30 lbs.	Sieve, organic impurities, specific gravity
	Cement	1 per 1000 cu. yds. Min. 1 per job	10 lbs.	◆ Fineness ◆ Chemical compound ◆ See mill reports
	Concrete Placement	1 per 50 cu. yds. Or each days pour (if less)		◆ Slump ◆ Air Test ◆ 5 compressive cylinder tests, test 2 at 7 days, 2 at 28 days, 1 hold
HMAC Surface Course	HMAC	1 per 500 tons or each days laydown		◆ 3 cores for density ◆ Extraction/gradation tests ◆ Stability tests ◆ Thickness ◆ Temperature
Pier or Footing Inspection	Inspection and verification of bearing	Each Pier or Slab Footing		Qualified Inspector with Engineer's Review
	Concrete & Steel Placement	Each Pier or Slab Footing		Qualified Inspector
	Inspection of Reinforcing	Slab Pre-pour and Cable Stressing		Qualified Inspector

-18-

MLA LABS, INC.   2804 LONGHORN BLVD.   AUSTIN, TEXAS 78758   512/873-8899   FAX 512/835-5114

█████ Elementary School
Engineer's Job No. █████

REFERENCES

1. Local geologic maps published by The Bureau of Economic Geology. Austin, Texas including:
   "Geologic Atlas of Texas" 15-minute quadrangles. March 9, 2004 geospatial data.
   "Geologic Map of the Austin Area, Texas 1992" Geology of Austin Area Plate VII.

2. The Geology of Texas, Volume I, Stratigraphy, The University of Texas Bulletin No. 3232: August 22, 1932, The University of Texas, Austin, Texas, 1981.

3. "Method for Determining Potential Vertical Rise, PVR, Test Method Tex-124-E", Manual of Testing Procedures, Texas Department of Transportation Materials and Tests Division, September 1995.

4. International Building Code 2003. International Code Council. 2003.

5. "ACI Committee Report 330R - Guide for Design and Construction of Concrete Parking Lots", ACI Manual of Concrete Practice - Part 2, American Concrete Institute, Farmington Hills, MI; 1996.

6. Design and Construction of Joints for Concrete Streets, Portland Cement Association, Arlington Heights, Illinois, 1992.

-19-

██████████ Elementary School
Engineer's Job No. ██████████

LIMITATIONS OF REPORT

The conditions of the site at locations other than the boring locations are not expressed or implied and conditions may be different at different times from the time of borings. Contractors or others desiring more information are advised to secure their own supplemental borings. This investigation and report, do not, and are not intended to determine the environmental conditions or evaluate possible hazardous or toxic waste conditions on this site or adjacent sites. Interested persons requiring this information are advised to contact MLA Labs, Inc.

The recommendations in this report are not intended to address the interior environmental effects of moisture migration through slabs. The Client is responsible for addressing the requirements of this project with respect to moisture migration through slab on ground foundations.

Diverse strategies can be applied during building design, construction, operation, and maintenance to prevent significant amounts of mold from growing on indoor surfaces. The geotechnical engineer in charge of this project is not a mold prevention consultant and none of the services performed in connection with this study were designed or conducted for the purpose of mold prevention. Proper implementation of the recommendations conveyed in this report may not of itself be sufficient to prevent mold from growing in or on the structure(s) involved.

The analysis and recommendations contained herein are based on the available data as shown in this report and the writer's professional expertise, experience and training, and no other warranty is expressed or implied concerning the satisfactory use of these recommendations or data.

██████████████

██████████████████████

-20-

APPENDIX A

GEOTECHNICAL DATA

MLA LABS, INC.   2804 LONGHORN BLVD.   AUSTIN, TEXAS 78758   512/873-8899   FAX 512/835-5114

Site in Yellow.

NAPP Aerial Photograph of Site – 1995

Source: TEXAS NATURAL RESOURCES INFORMATION SYSTEM
3.75-minute DOQQ. 1-meter ground resolution. apx. date 1995-6
(http://www.tnris.state.tx.us/digital.htm)

\\File_server\mlalabs\GeoActive███████████Elementary School███████Site Maps.doc

MLA LABS, INC.   2804 LONGHORN BLVD.   AUSTIN, TEXAS 78758   512/873-8899   FAX 512/835-5114

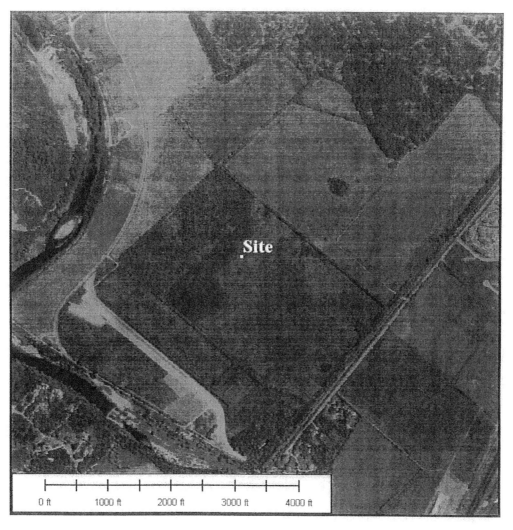

Site in Yellow.

NAIP Aerial Photograph of Site – 2004

Source: TEXAS NATURAL RESOURCES INFORMATION SYSTEM
3.75-minute DOQQ. 1-meter ground resolution. apx. date 2004
(http://www.tnris.state.tx.us)

\\File_server\mlalabs\GeoActive ████████████ Elementary School ██████████ Maps.doc

MLA LABS, INC.   2804 LONGHORN BLVD.   AUSTIN, TEXAS 78758   512/873-8899   FAX 512/835-5114

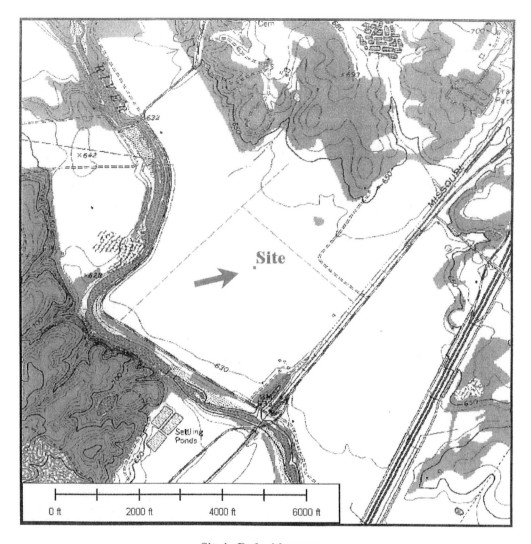

Site in Red with arrow.

U.S. 7.5 Minute Series Topographic Map
San Marcos North Quadrangle, Texas
Contour Interval = 10 feet
Source: TEXAS NATURAL RESOURCES INFORMATION SYSTEM
(http://www.tnris.state.tx.us/digital.htm)

\\File_server\mlalabs\GeoActive\▮▮▮▮▮Elementary▮▮▮▮▮Site Maps.doc

MLA LABS, INC.   2804 LONGHORN BLVD.   AUSTIN, TEXAS 78758   512/873-8899   FAX 512/835-5114

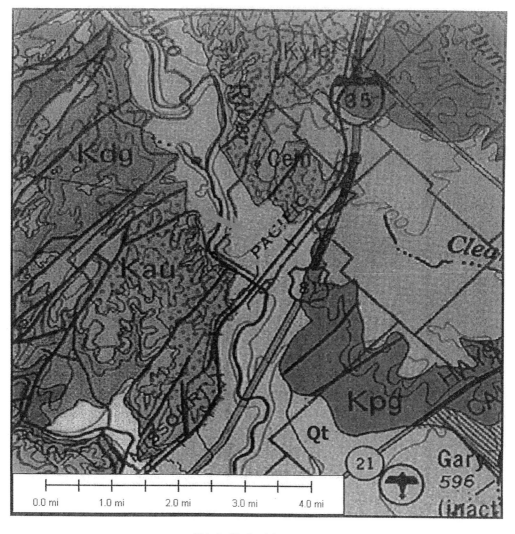

Site in Red with arrow.

Geologic Setting of Site
Geologic Atlas of Texas
Contour Interval = 50 feet
Original Source: Bureau of Economic Geology, The University of Texas at Austin, latest version
Digital Source: 15-minute Digital GAT Quads. TCEQ March 9, 2004

\\File_server\mlalabs\GeoActive█████████Elementary███████Site Maps.doc

MLA LABS, INC.   2804 LONGHORN BLVD.   AUSTIN, TEXAS 78758   512/873-8899   FAX 512/835-5114

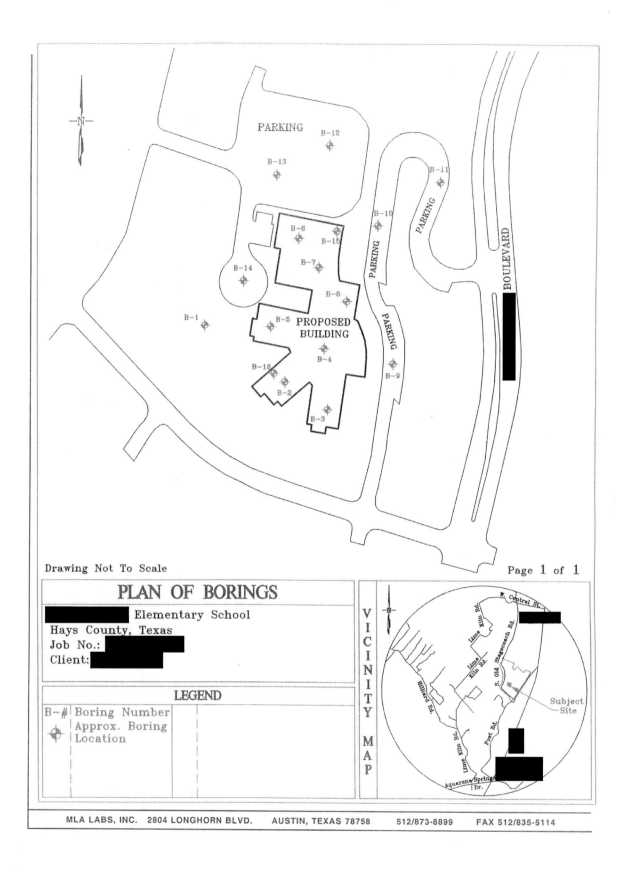

Drawing Not To Scale

Page 1 of 1

PLAN OF BORINGS

� Elementary School
Hays County, Texas
Job No.:
Client:

LEGEND

B-#	Boring Number		
⊕	Approx. Boring Location		

VICINITY MAP

MLA LABS, INC.   2804 LONGHORN BLVD.   AUSTIN, TEXAS 78758   512/873-8899   FAX 512/835-5114

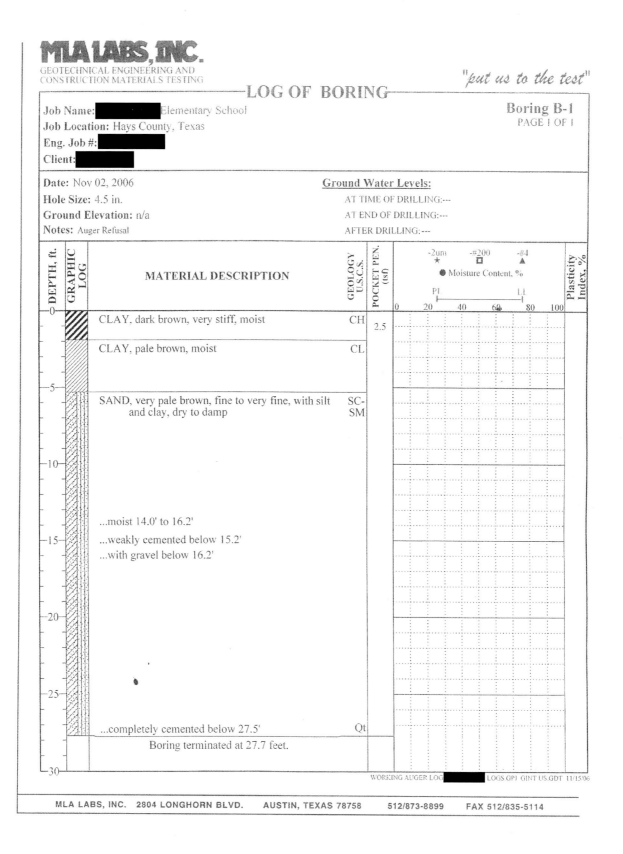

MLA LABS, INC.
GEOTECHNICAL ENGINEERING AND
CONSTRUCTION MATERIALS TESTING

"put us to the test"

LOG OF BORING

Job Name: ▮▮▮▮▮ Elementary School
Job Location: Hays County, Texas
Eng. Job #: ▮▮▮▮▮▮
Client: ▮▮▮▮▮

Boring B-1
PAGE 1 OF 1

Date: Nov 02, 2006
Hole Size: 4.5 in.
Ground Elevation: n/a
Notes: Auger Refusal

Ground Water Levels:
AT TIME OF DRILLING:---
AT END OF DRILLING:---
AFTER DRILLING:---

DEPTH, ft.	GRAPHIC LOG	MATERIAL DESCRIPTION	GEOLOGY U.S.C.S.	POCKET PEN. (tsf)
0		CLAY, dark brown, very stiff, moist	CH	2.5
		CLAY, pale brown, moist	CL	
5		SAND, very pale brown, fine to very fine, with silt and clay, dry to damp	SC-SM	
10				
		...moist 14.0' to 16.2'		
15		...weakly cemented below 15.2'		
		...with gravel below 16.2'		
20				
25				
		...completely cemented below 27.5'	Qt	
		Boring terminated at 27.7 feet.		
30				

Moisture Content, %
-2um ★   -#200 □   -#4 ▲
PI — LL
0 20 40 60 80 100

Plasticity Index, %

WORKING AUGER LOG ▮▮▮▮▮ LOGS.GPJ GINT US.GDT 11/15/06

MLA LABS, INC.   2804 LONGHORN BLVD.   AUSTIN, TEXAS 78758   512/873-8899   FAX 512/835-5114

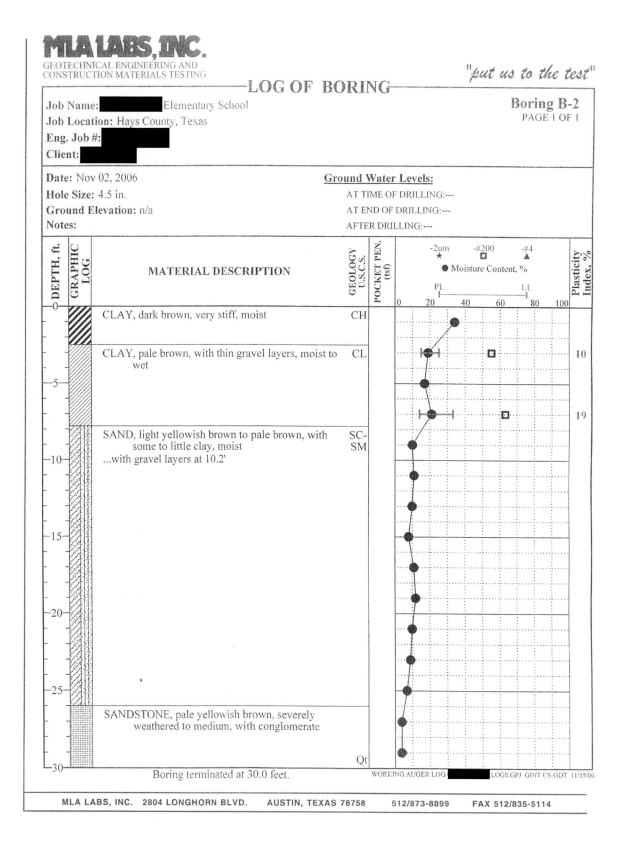

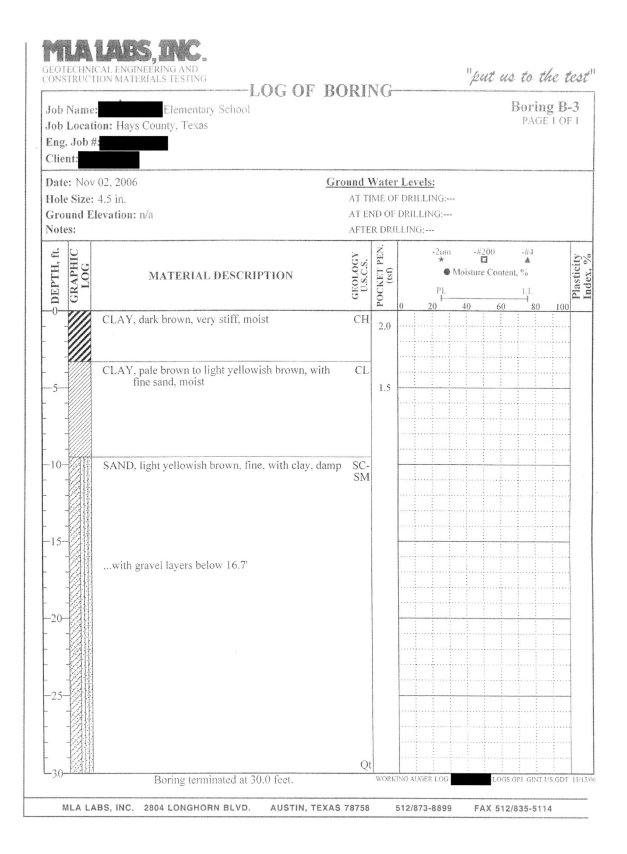

MLA LABS, INC.
GEOTECHNICAL ENGINEERING AND
CONSTRUCTION MATERIALS TESTING

"put us to the test"

LOG OF BORING

Boring B-3
PAGE 1 OF 1

Job Name: ▮▮▮▮▮ Elementary School
Job Location: Hays County, Texas
Eng. Job #: ▮▮▮▮▮
Client: ▮▮▮▮▮

Date: Nov 02, 2006
Hole Size: 4.5 in.
Ground Elevation: n/a
Notes:

Ground Water Levels:
AT TIME OF DRILLING:---
AT END OF DRILLING:---
AFTER DRILLING:---

DEPTH, ft.	GRAPHIC LOG	MATERIAL DESCRIPTION	GEOLOGY U.S.C.S.	POCKET PEN. (tsf)	Plot	Plasticity Index, %
0		CLAY, dark brown, very stiff, moist	CH	2.0		
5		CLAY, pale brown to light yellowish brown, with fine sand, moist	CL	1.5		
10		SAND, light yellowish brown, fine, with clay, damp	SC-SM			
15		...with gravel layers below 16.7'				
20						
25						
30			Qt			

Plot legend: -2um ★   -#200 ☐   -#4 ▲   ● Moisture Content, %   PL.———LL.   0 20 40 60 80 100

Boring terminated at 30.0 feet.

WORKING AUGER LOG ▮▮▮▮▮ LOGS.GPJ GINT US.GDT 11/15/06

MLA LABS, INC.   2804 LONGHORN BLVD.   AUSTIN, TEXAS 78758   512/873-8899   FAX 512/835-5114

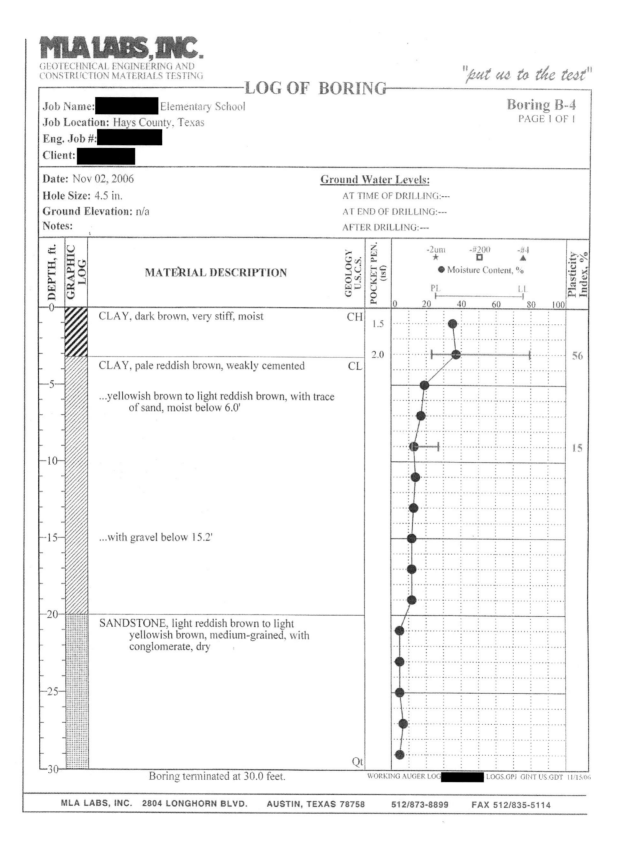

MLA LABS, INC.
GEOTECHNICAL ENGINEERING AND
CONSTRUCTION MATERIALS TESTING

"put us to the test"

LOG OF BORING

Job Name: ▮▮▮▮ Elementary School
Job Location: Hays County, Texas
Eng. Job #: ▮▮▮▮
Client: ▮▮▮▮

Boring B-4
PAGE 1 OF 1

Date: Nov 02, 2006
Hole Size: 4.5 in.
Ground Elevation: n/a
Notes:

Ground Water Levels:
AT TIME OF DRILLING:---
AT END OF DRILLING:---
AFTER DRILLING:---

DEPTH, ft.	GRAPHIC LOG	MATERIAL DESCRIPTION	GEOLOGY U.S.C.S.	POCKET PEN. (tsf)	Moisture Content, %	Plasticity Index, %
0		CLAY, dark brown, very stiff, moist	CH	1.5		
				2.0		56
5		CLAY, pale reddish brown, weakly cemented	CL			
		...yellowish brown to light reddish brown, with trace of sand, moist below 6.0'				
10						15
15		...with gravel below 15.2'				
20		SANDSTONE, light reddish brown to light yellowish brown, medium-grained, with conglomerate, dry				
25						
30			Qt			

Boring terminated at 30.0 feet.

WORKING AUGER LOG ▮▮▮▮ LOGS.GPJ GINT US.GDT 11/15/06

MLA LABS, INC.   2804 LONGHORN BLVD.   AUSTIN, TEXAS 78758   512/873-8899   FAX 512/835-5114

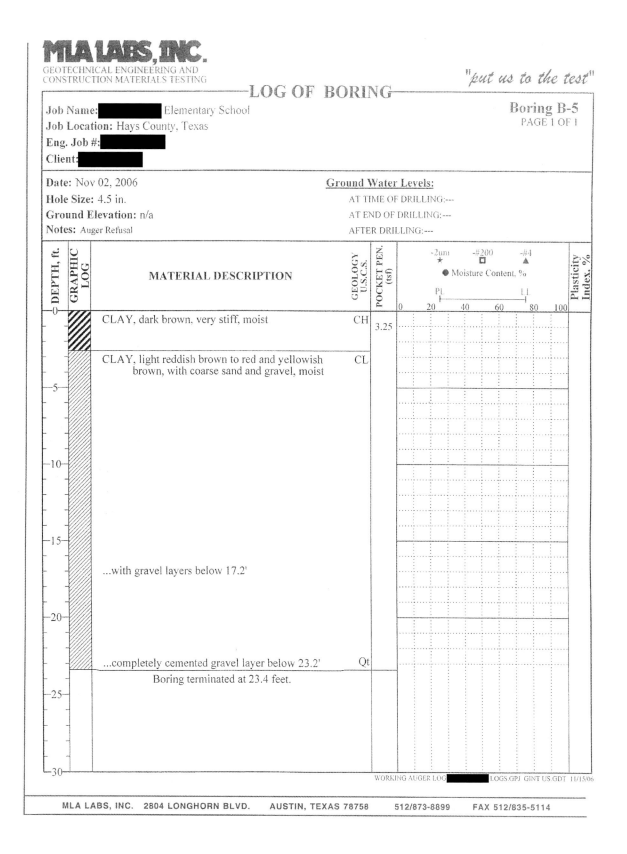

MLA LABS, INC.

GEOTECHNICAL ENGINEERING AND
CONSTRUCTION MATERIALS TESTING

"put us to the test"

—LOG OF BORING—

Job Name: ▓▓▓▓ Elementary School

Job Location: Hays County, Texas

Eng. Job #: ▓▓▓▓▓

Client: ▓▓▓▓

Boring B-5

PAGE 1 OF 1

Date: Nov 02, 2006

Hole Size: 4.5 in.

Ground Elevation: n/a

Notes: Auger Refusal

Ground Water Levels:

AT TIME OF DRILLING:---

AT END OF DRILLING:---

AFTER DRILLING:---

DEPTH, ft.	GRAPHIC LOG	MATERIAL DESCRIPTION	GEOLOGY U.S.C.S.	POCKET PEN. (tsf)
0		CLAY, dark brown, very stiff, moist	CH	3.25
		CLAY, light reddish brown to red and yellowish brown, with coarse sand and gravel, moist	CL	
5				
10				
15				
		...with gravel layers below 17.2'		
20				
		...completely cemented gravel layer below 23.2'	Qt	
		Boring terminated at 23.4 feet.		
25				
30				

Moisture Content, %
-2um ★ -#200 □ -#4 ▲
PL ├──────┤ LL
0 20 40 60 80 100
Plasticity Index, %

WORKING AUGER LOG ▓▓▓▓ LOGS.GPJ GINT US.GDT 11/15/06

MLA LABS, INC.   2804 LONGHORN BLVD.   AUSTIN, TEXAS 78758   512/873-8899   FAX 512/835-5114

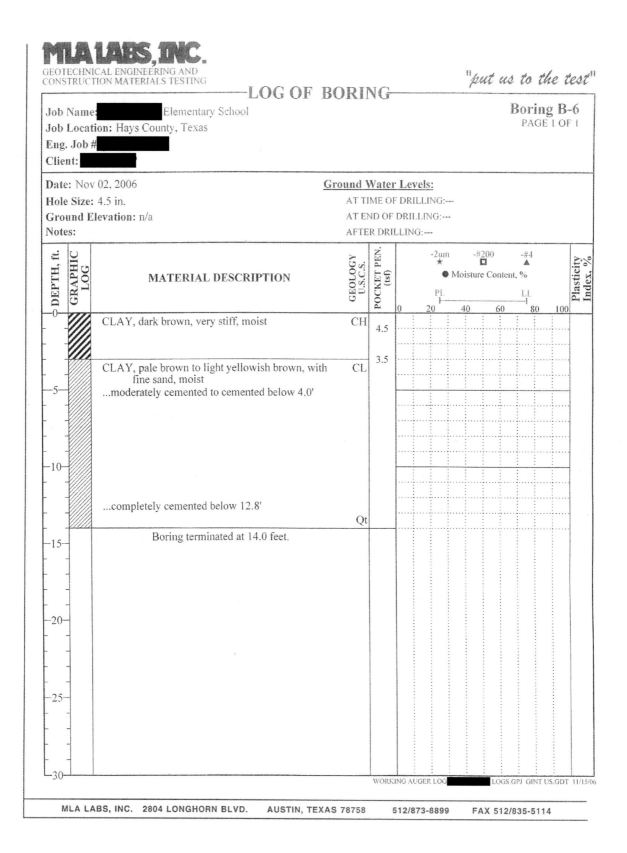

MLA LABS, INC.
GEOTECHNICAL ENGINEERING AND
CONSTRUCTION MATERIALS TESTING

"put us to the test"

LOG OF BORING

Boring B-6
PAGE 1 OF 1

Job Name: ▮▮▮▮ Elementary School
Job Location: Hays County, Texas
Eng. Job # ▮▮▮▮
Client: ▮▮▮▮

Date: Nov 02, 2006
Hole Size: 4.5 in.
Ground Elevation: n/a
Notes:

Ground Water Levels:
AT TIME OF DRILLING:---
AT END OF DRILLING:---
AFTER DRILLING:---

DEPTH, ft.	GRAPHIC LOG	MATERIAL DESCRIPTION	GEOLOGY U.S.C.S.	POCKET PEN. (tsf)	Moisture Content, %	Plasticity Index, %
0		CLAY, dark brown, very stiff, moist	CH	4.5		
		CLAY, pale brown to light yellowish brown, with fine sand, moist ...moderately cemented to cemented below 4.0'	CL	3.5		
5						
10						
		...completely cemented below 12.8'	Qt			
15		Boring terminated at 14.0 feet.				
20						
25						
30						

WORKING AUGER LOG ▮▮▮▮ LOGS.GPJ GINT US.GDT 11/15/06

MLA LABS, INC.   2804 LONGHORN BLVD.   AUSTIN, TEXAS 78758   512/873-8899   FAX 512/835-5114

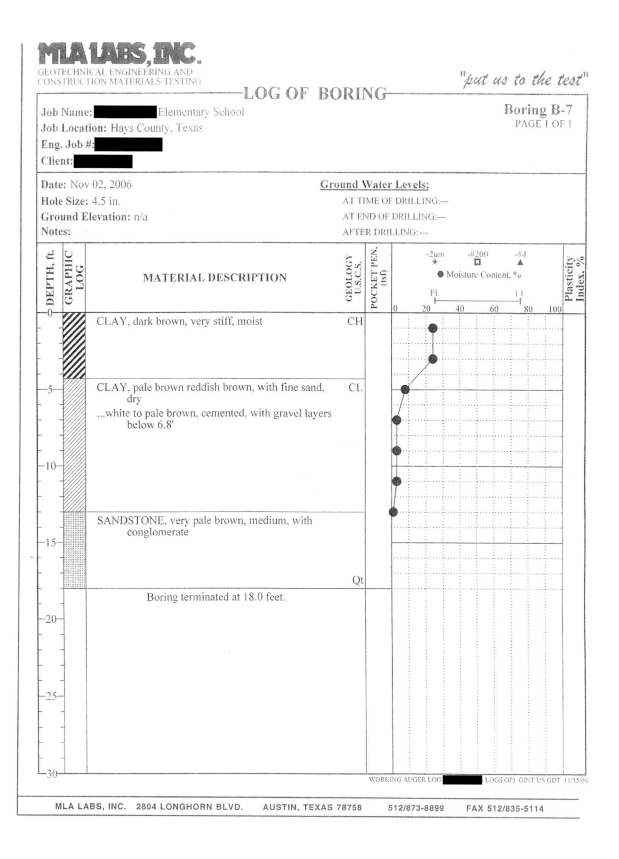

MLA LABS, INC.
GEOTECHNICAL ENGINEERING AND
CONSTRUCTION MATERIALS TESTING

"put us to the test"

LOG OF BORING

Job Name: ▮▮▮▮▮ Elementary School
Job Location: Hays County, Texas
Eng. Job #: ▮▮▮▮▮
Client: ▮▮▮▮▮

Boring B-7
PAGE 1 OF 1

Date: Nov 02, 2006
Hole Size: 4.5 in.
Ground Elevation: n/a
Notes:

Ground Water Levels:
AT TIME OF DRILLING:---
AT END OF DRILLING:---
AFTER DRILLING:---

DEPTH, ft.	GRAPHIC LOG	MATERIAL DESCRIPTION	GEOLOGY U.S.C.S.	POCKET PEN. (tsf)	Moisture Content, %	Plasticity Index, %
0		CLAY, dark brown, very stiff, moist	CH			
5		CLAY, pale brown reddish brown, with fine sand, dry ...white to pale brown, cemented, with gravel layers below 6.8'	CL			
10						
15		SANDSTONE, very pale brown, medium, with conglomerate				
			Qt			
		Boring terminated at 18.0 feet.				
20						
25						
30						

-2um ★   -#200 □   -#4 ▲
● Moisture Content, %
PL   LL
0 20 40 60 80 100

WORKING AUGER LOG ▮▮▮▮▮ LOGS.GPJ GINT US.GDT 11/15/06

MLA LABS, INC.   2804 LONGHORN BLVD.   AUSTIN, TEXAS 78758   512/873-8899   FAX 512/835-5114

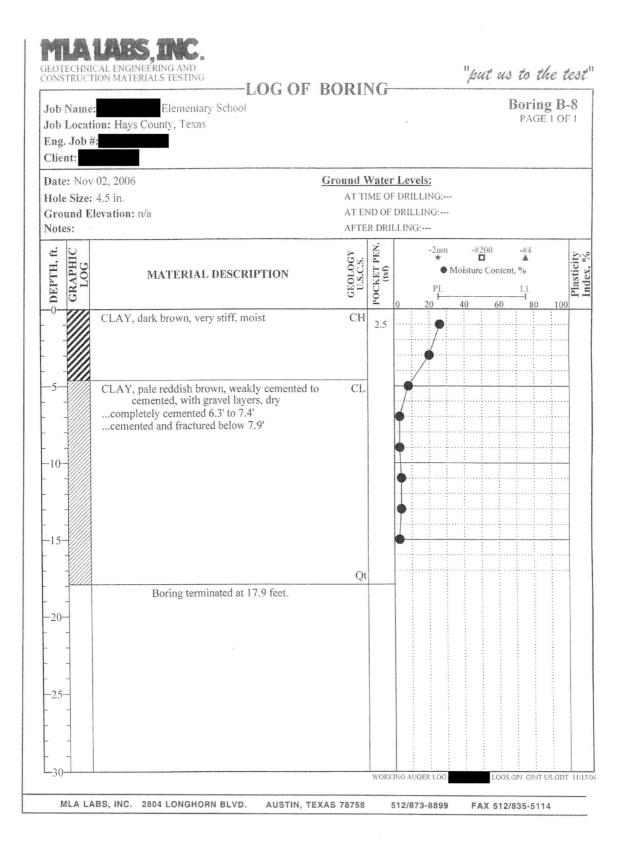

MLA LABS, INC.
GEOTECHNICAL ENGINEERING AND
CONSTRUCTION MATERIALS TESTING

LOG OF BORING

"put us to the test"

Job Name: ▮▮▮▮▮ Elementary School
Job Location: Hays County, Texas
Eng. Job #: ▮▮▮▮▮
Client: ▮▮▮▮

Boring B-8
PAGE 1 OF 1

Date: Nov 02, 2006
Hole Size: 4.5 in.
Ground Elevation: n/a
Notes:

Ground Water Levels:
AT TIME OF DRILLING:---
AT END OF DRILLING:---
AFTER DRILLING:---

DEPTH, ft.	GRAPHIC LOG	MATERIAL DESCRIPTION	GEOLOGY U.S.C.S.	POCKET PEN. (tsf)
0		CLAY, dark brown, very stiff, moist	CH	2.5
5		CLAY, pale reddish brown, weakly cemented to cemented, with gravel layers, dry ...completely cemented 6.3' to 7.4' ...cemented and fractured below 7.9'	CL	
10				
15				
			Qt	
		Boring terminated at 17.9 feet.		
20				
25				
30				

-2um ★   -#200 □   -#4 ▲
● Moisture Content, %
PL   LL
0 20 40 60 80 100

Plasticity Index, %

WORKING AUGER LOG ▮▮▮▮ LOGS.GPJ GINT US.GDT 11/15/06

MLA LABS, INC.   2804 LONGHORN BLVD.   AUSTIN, TEXAS 78758   512/873-8899   FAX 512/835-5114

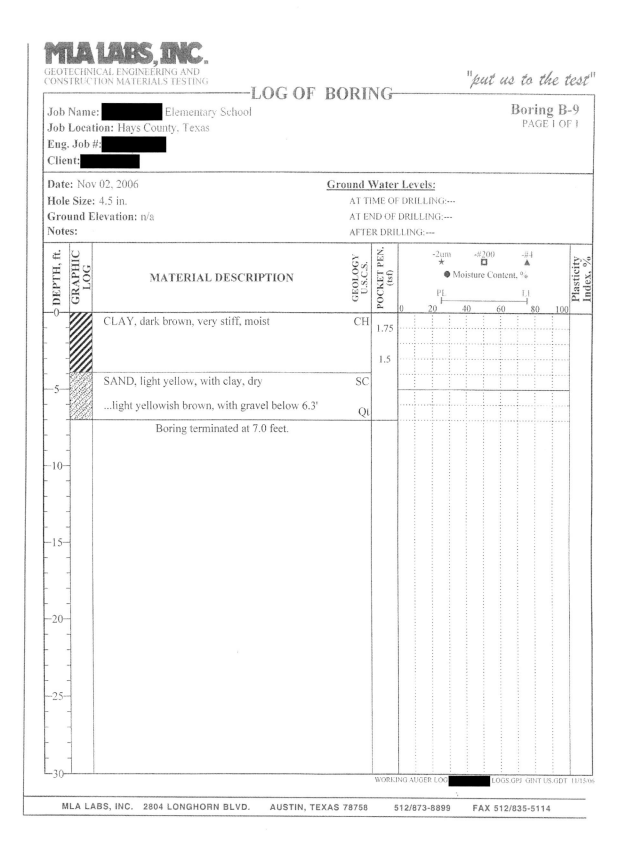

MLA LABS, INC.

GEOTECHNICAL ENGINEERING AND
CONSTRUCTION MATERIALS TESTING

"put us to the test"

LOG OF BORING

Job Name: ▮▮▮ Elementary School

Job Location: Hays County, Texas

Eng. Job #: ▮▮▮▮

Client: ▮▮▮

Boring B-9

PAGE 1 OF 1

Date: Nov 02, 2006

Hole Size: 4.5 in.

Ground Elevation: n/a

Notes:

Ground Water Levels:

AT TIME OF DRILLING:---

AT END OF DRILLING:---

AFTER DRILLING:---

DEPTH, ft.	GRAPHIC LOG	MATERIAL DESCRIPTION	GEOLOGY U.S.C.S.	POCKET PEN. (tsf)
0		CLAY, dark brown, very stiff, moist	CH	1.75
				1.5
5		SAND, light yellow, with clay, dry	SC	
		...light yellowish brown, with gravel below 6.3'	Qt	
		Boring terminated at 7.0 feet.		

Chart legend: -2um ★ -#200 □ -#4 ▲ ● Moisture Content, % PL — LL (scale 0 20 40 60 80 100) Plasticity Index, %

WORKING AUGER LOG ▮▮▮ LOGS.GPJ GINT US.GDT 11/15/06

MLA LABS, INC.   2804 LONGHORN BLVD.   AUSTIN, TEXAS 78758   512/873-8899   FAX 512/835-5114

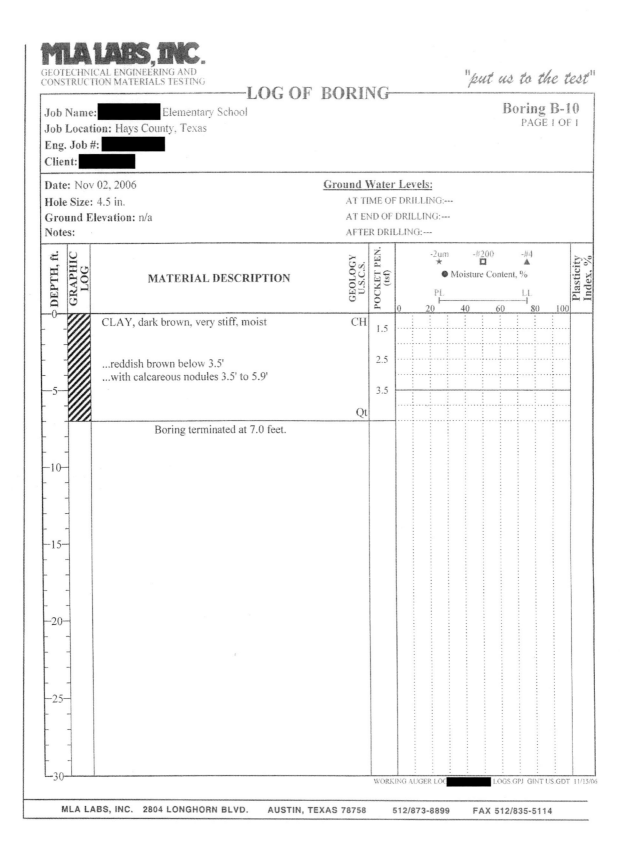

MLA LABS, INC.
GEOTECHNICAL ENGINEERING AND
CONSTRUCTION MATERIALS TESTING

"put us to the test"

LOG OF BORING

Job Name: ████ Elementary School
Job Location: Hays County, Texas
Eng. Job #: ████
Client: ████

Boring B-10
PAGE 1 OF 1

Date: Nov 02, 2006
Hole Size: 4.5 in.
Ground Elevation: n/a
Notes:

Ground Water Levels:
AT TIME OF DRILLING:---
AT END OF DRILLING:---
AFTER DRILLING:---

DEPTH, ft.	GRAPHIC LOG	MATERIAL DESCRIPTION	GEOLOGY U.S.C.S.	POCKET PEN. (tsf)
0		CLAY, dark brown, very stiff, moist	CH	1.5
				2.5
		...reddish brown below 3.5'		
		...with calcareous nodules 3.5' to 5.9'		
5				3.5
			Qt	
		Boring terminated at 7.0 feet.		

-2um ★   -#200 ☐   -#4 ▲
● Moisture Content, %
PL ⊢———⊣ LL
0 20 40 60 80 100

Plasticity Index, %

WORKING AUGER LOG ████ LOGS.GPJ GINT US.GDT 11/15/06

MLA LABS, INC.   2804 LONGHORN BLVD.   AUSTIN, TEXAS 78758   512/873-8899   FAX 512/835-5114

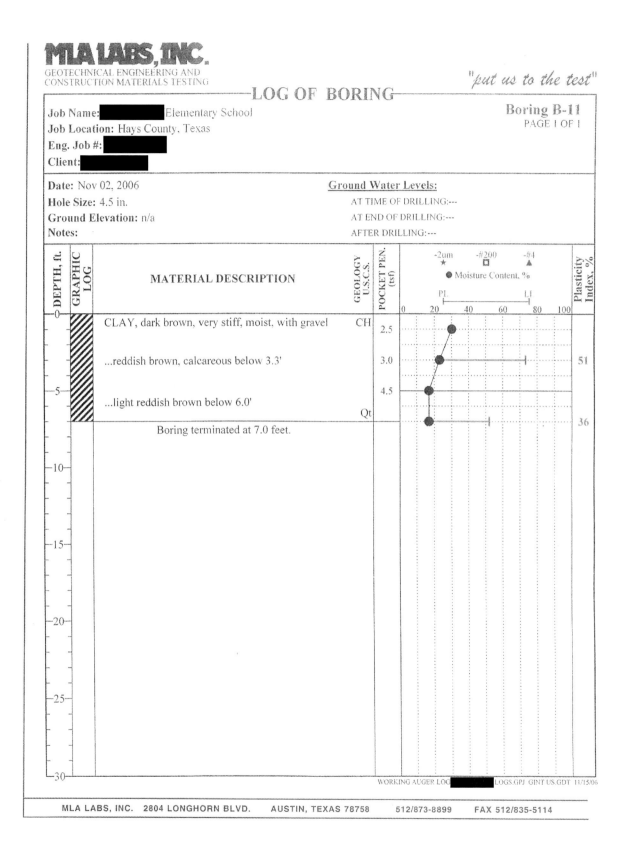

MLA LABS, INC.
GEOTECHNICAL ENGINEERING AND
CONSTRUCTION MATERIALS TESTING

"put us to the test"

LOG OF BORING

Boring B-11
PAGE 1 OF 1

Job Name: ▮▮▮▮ Elementary School
Job Location: Hays County, Texas
Eng. Job #: ▮▮▮▮
Client: ▮▮▮▮

Date: Nov 02, 2006
Hole Size: 4.5 in.
Ground Elevation: n/a
Notes:

Ground Water Levels:
AT TIME OF DRILLING:---
AT END OF DRILLING:---
AFTER DRILLING:---

DEPTH, ft.	GRAPHIC LOG	MATERIAL DESCRIPTION	GEOLOGY U.S.C.S.	POCKET PEN. (tsf)	Moisture Content, % / PL — LI	Plasticity Index, %
0		CLAY, dark brown, very stiff, moist, with gravel	CH	2.5		
		...reddish brown, calcareous below 3.3'		3.0		51
5		...light reddish brown below 6.0'		4.5		
			Qt			36
		Boring terminated at 7.0 feet.				

WORKING AUGER LOG ▮▮▮▮ LOGS.GPJ GINT US.GDT 11/15/06

MLA LABS, INC.   2804 LONGHORN BLVD.   AUSTIN, TEXAS 78758   512/873-8899   FAX 512/835-5114

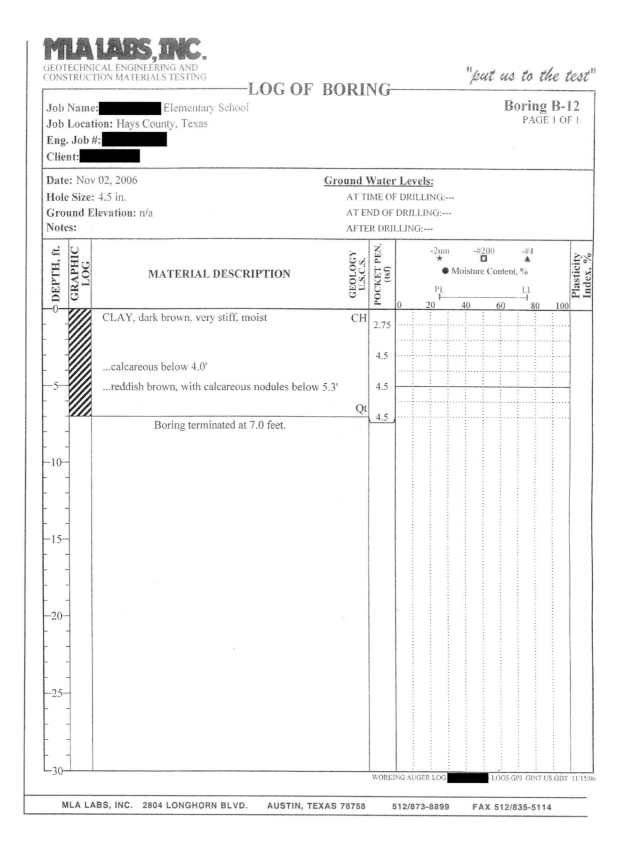

MLA LABS, INC.
GEOTECHNICAL ENGINEERING AND
CONSTRUCTION MATERIALS TESTING

LOG OF BORING

"put us to the test"

Job Name: ▇▇▇▇ Elementary School
Job Location: Hays County. Texas
Eng. Job #: ▇▇▇▇
Client: ▇▇▇

Boring B-12
PAGE 1 OF 1

Date: Nov 02, 2006
Hole Size: 4.5 in.
Ground Elevation: n/a
Notes:

Ground Water Levels:
AT TIME OF DRILLING:---
AT END OF DRILLING:---
AFTER DRILLING:---

DEPTH, ft.	GRAPHIC LOG	MATERIAL DESCRIPTION	GEOLOGY U.S.C.S.	POCKET PEN. (tsf)
0		CLAY, dark brown, very stiff, moist	CH	2.75
				4.5
		...calcareous below 4.0'		4.5
5		...reddish brown, with calcareous nodules below 5.3'		4.5
			Qt	4.5
		Boring terminated at 7.0 feet.		

-2um ★   -#200 ▢   -#4 ▲
● Moisture Content, %
PL |———| LL
0   20   40   60   80   100

Plasticity Index, %

WORKING AUGER LOG ▇▇▇ LOGS.GPJ GINT US.GDT 11/15/06

MLA LABS, INC.   2804 LONGHORN BLVD.   AUSTIN, TEXAS 78758   512/873-8899   FAX 512/835-5114

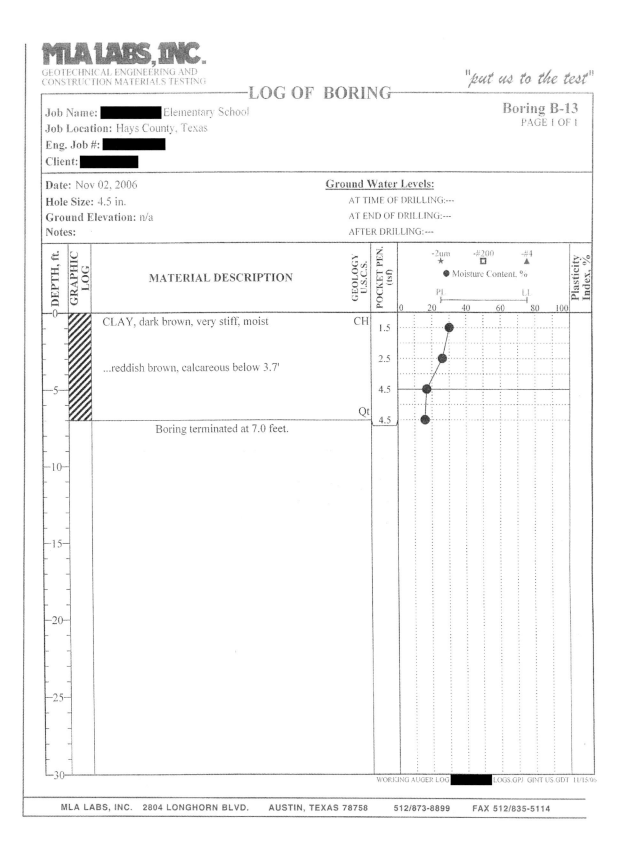

MLA LABS, INC.
GEOTECHNICAL ENGINEERING AND
CONSTRUCTION MATERIALS TESTING

"put us to the test"

LOG OF BORING

Job Name: ▮▮▮▮ Elementary School

Job Location: Hays County, Texas

Eng. Job #: ▮▮▮▮

Client: ▮▮▮▮

Boring B-13
PAGE 1 OF 1

Date: Nov 02, 2006

Hole Size: 4.5 in.

Ground Elevation: n/a

Notes:

Ground Water Levels:

AT TIME OF DRILLING:---

AT END OF DRILLING:---

AFTER DRILLING:---

DEPTH, ft.	GRAPHIC LOG	MATERIAL DESCRIPTION	GEOLOGY U.S.C.S.	POCKET PEN. (tsf)	Moisture Content, %	Plasticity Index, %
0		CLAY, dark brown, very stiff, moist	CH	1.5		
		...reddish brown, calcareous below 3.7'		2.5		
5				4.5		
			Qt	4.5		
		Boring terminated at 7.0 feet.				

-2um ★   -#200 ☐   -#4 ▲
● Moisture Content, %
PL |——————| LL
0   20   40   60   80   100

MLA LABS, INC.   2804 LONGHORN BLVD.   AUSTIN, TEXAS 78758   512/873-8899   FAX 512/835-5114

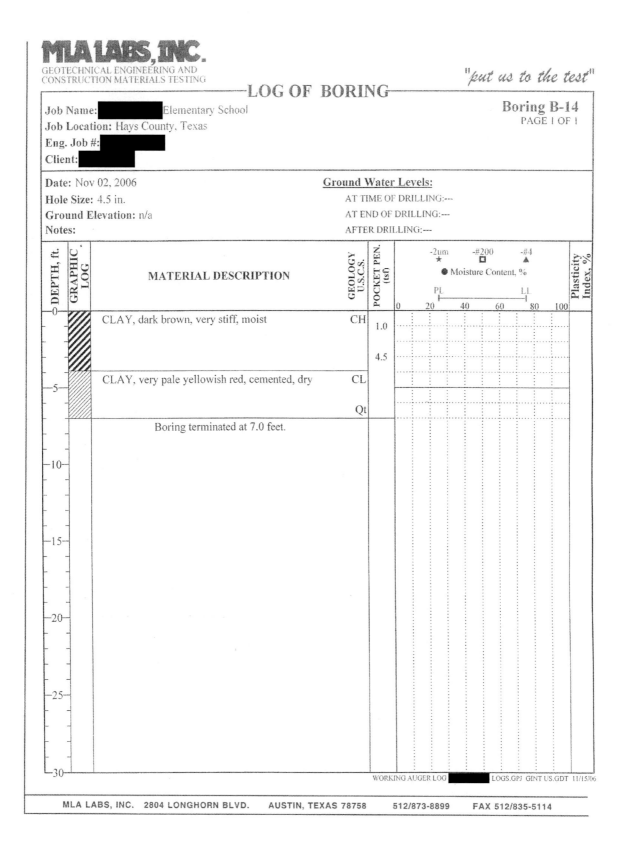

MLA LABS, INC.
GEOTECHNICAL ENGINEERING AND
CONSTRUCTION MATERIALS TESTING

"put us to the test"

LOG OF BORING

Boring B-14
PAGE 1 OF 1

Job Name: ▓▓▓ Elementary School
Job Location: Hays County, Texas
Eng. Job #: ▓▓▓
Client: ▓▓▓

Date: Nov 02, 2006
Hole Size: 4.5 in.
Ground Elevation: n/a
Notes:

Ground Water Levels:
AT TIME OF DRILLING:---
AT END OF DRILLING:---
AFTER DRILLING:---

DEPTH, ft.	GRAPHIC LOG	MATERIAL DESCRIPTION	GEOLOGY U.S.C.S.	POCKET PEN. (tsf)
0		CLAY, dark brown, very stiff, moist	CH	1.0
				4.5
5		CLAY, very pale yellowish red, cemented, dry	CL	
			Qt	
		Boring terminated at 7.0 feet.		

Plot legend: -2um ★   -#200 ▢   -#4 ▲   ● Moisture Content, % ; PL / LL scale 0 20 40 60 80 100 ; Plasticity Index, %

WORKING AUGER LOG ▓▓▓ LOGS.GPJ GINT US.GDT 11/15/06

MLA LABS, INC.   2804 LONGHORN BLVD.   AUSTIN, TEXAS 78758   512/873-8899   FAX 512/835-5114

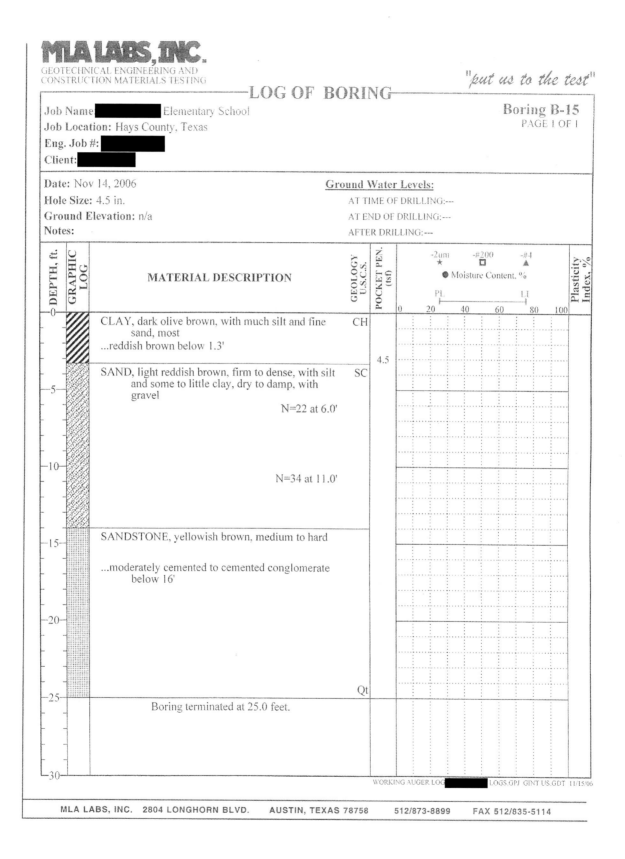

MLA LABS, INC.
GEOTECHNICAL ENGINEERING AND
CONSTRUCTION MATERIALS TESTING

LOG OF BORING

"put us to the test"

Job Name: ▮▮▮▮ Elementary School

Job Location: Hays County, Texas

Eng. Job #: ▮▮▮▮

Client: ▮▮▮▮

Boring B-15

PAGE 1 OF 1

Date: Nov 14, 2006

Hole Size: 4.5 in.

Ground Elevation: n/a

Notes:

Ground Water Levels:

AT TIME OF DRILLING:---

AT END OF DRILLING:---

AFTER DRILLING:---

DEPTH, ft.	GRAPHIC LOG	MATERIAL DESCRIPTION	GEOLOGY U.S.C.S.	POCKET PEN. (tsf)
0		CLAY, dark olive brown, with much silt and fine sand, most ...reddish brown below 1.3'	CH	4.5
5		SAND, light reddish brown, firm to dense, with silt and some to little clay, dry to damp, with gravel N=22 at 6.0'	SC	
10		N=34 at 11.0'		
15		SANDSTONE, yellowish brown, medium to hard ...moderately cemented to cemented conglomerate below 16'		
20				
25			Qt	
		Boring terminated at 25.0 feet.		
30				

Graph legend: -2μm ★ -#200 ◻ -#4 ▲ ● Moisture Content, % PL — LI Plasticity Index, % (scale 0 20 40 60 80 100)

WORKING AUGER LOG ▮▮▮▮ LOGS.GPJ GINT US.GDT 11/15/06

MLA LABS, INC.   2804 LONGHORN BLVD.   AUSTIN, TEXAS 78758   512/873-8899   FAX 512/835-5114

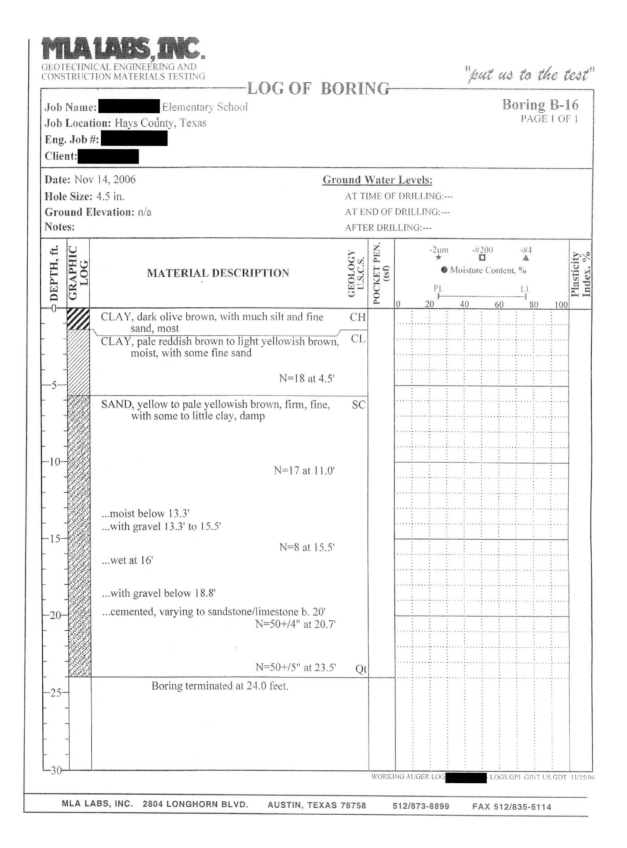

MLA LABS, INC.
GEOTECHNICAL ENGINEERING AND
CONSTRUCTION MATERIALS TESTING

"put us to the test"

LOG OF BORING

Boring B-16
PAGE 1 OF 1

Job Name: ▮▮▮▮ Elementary School
Job Location: Hays County, Texas
Eng. Job #: ▮▮▮▮
Client: ▮▮▮▮

Date: Nov 14, 2006
Hole Size: 4.5 in.
Ground Elevation: n/a
Notes:

Ground Water Levels:
AT TIME OF DRILLING:---
AT END OF DRILLING:---
AFTER DRILLING:---

DEPTH, ft.	GRAPHIC LOG	MATERIAL DESCRIPTION	GEOLOGY U.S.C.S.	POCKET PEN. (tsf)	Plasticity Index, %
0		CLAY, dark olive brown, with much silt and fine sand, most	CH		
		CLAY, pale reddish brown to light yellowish brown, moist, with some fine sand	CL		
5		N=18 at 4.5'			
		SAND, yellow to pale yellowish brown, firm, fine, with some to little clay, damp	SC		
10		N=17 at 11.0'			
15		...moist below 13.3' ...with gravel 13.3' to 15.5' N=8 at 15.5' ...wet at 16'			
20		...with gravel below 18.8' ...cemented, varying to sandstone/limestone b. 20' N=50+/4" at 20.7'			
		N=50+/5" at 23.5'	Qt		
25		Boring terminated at 24.0 feet.			
30					

Moisture Content, %
-2um ★   -#200 □   -#4 ▲
● Moisture Content, %
PL |—————————| LL
0   20   40   60   80   100

WORKING AUGER LOG ▮▮▮▮-LOGS.GPJ GINT US.GDT 11/15/06

MLA LABS, INC.   2804 LONGHORN BLVD.   AUSTIN, TEXAS 78758   512/873-8899   FAX 512/835-5114

SOIL CLASSIFICATION CHART

MAJOR DIVISIONS			SYMBOLS		TYPICAL DESCRIPTIONS
			GRAPH	LETTER	
COARSE GRAINED SOILS MORE THAN 50% OF MATERIAL IS LARGER THAN NO. 200 SIEVE SIZE	GRAVEL AND GRAVELLY SOILS MORE THAN 50% OF COARSE FRACTION RETAINED ON NO. 4 SIEVE	CLEAN GRAVELS		GW	WELL-GRADED GRAVELS, GRAVEL - SAND MIXTURES, LITTLE OR NO FINES
		(LITTLE OR NO FINES)		GP	POORLY-GRADED GRAVELS, GRAVEL - SAND MIXTURES, LITTLE OR NO FINES
		GRAVELS WITH FINES		GM	SILTY GRAVELS, GRAVEL - SAND - SILT MIXTURES
		(APPRECIABLE AMOUNT OF FINES)		GC	CLAYEY GRAVELS, GRAVEL - SAND - CLAY MIXTURES
	SAND AND SANDY SOILS MORE THAN 50% OF COARSE FRACTION PASSING ON NO. 4 SIEVE	CLEAN SANDS		SW	WELL-GRADED SANDS, GRAVELLY SANDS, LITTLE OR NO FINES
		(LITTLE OR NO FINES)		SP	POORLY-GRADED SANDS, GRAVELLY SAND, LITTLE OR NO FINES
		SANDS WITH FINES		SM	SILTY SANDS, SAND - SILT MIXTURES
		(APPRECIABLE AMOUNT OF FINES)		SC	CLAYEY SANDS, SAND - CLAY MIXTURES
FINE GRAINED SOILS MORE THAN 50% OF MATERIAL IS SMALLER THAN NO. 200 SIEVE SIZE	SILTS AND CLAYS	LIQUID LIMIT LESS THAN 50		ML	INORGANIC SILTS AND VERY FINE SANDS, ROCK FLOUR, SILTY OR CLAYEY FINE SANDS OR CLAYEY SILTS WITH SLIGHT PLASTICITY
				CL	INORGANIC CLAYS OF LOW TO MEDIUM PLASTICITY, GRAVELLY CLAYS, SANDY CLAYS, SILTY CLAYS, LEAN CLAYS
				OL	ORGANIC SILTS AND ORGANIC SILTY CLAYS OF LOW PLASTICITY
	SILTS AND CLAYS	LIQUID LIMIT GREATER THAN 50		MH	INORGANIC SILTS, MICACEOUS OR DIATOMACEOUS FINE SAND OR SILTY SOILS
				CH	INORGANIC CLAYS OF HIGH PLASTICITY
				OH	ORGANIC CLAYS OF MEDIUM TO HIGH PLASTICITY, ORGANIC SILTS
SOILS OF MODERATE PLASTICITY				CL-CH	LOW PI CLAYS WITH APPRECIABLE HIGH PI MOTTLING, CLAY WITH BORDERLINE CLASSIFICATION

NOTE: DUAL SYMBOLS ARE USED TO INDICATE BORDERLINE SOIL CLASSIFICATIONS

MLA LABS, INC.   2804 LONGHORN BLVD.   AUSTIN, TEXAS 78758   512/873-8899   FAX 512/835-5114

Key to Terms and Abbreviations

Descriptive Terms Characterizing Soils and Rock	Standard Description Abbreviations and Terms	Symbols and Abbreviations for Test Data
Slickensided – having inclined planes of weakness that are slick and glossy in appearance. **Fissured** – containing shrinkage cracks frequently filled with fine sand or silt, usually more or less vertical. **Laminated** – composed of thin layers of varying color or texture. Layers are typically distinct and varying in composition from sand to silt and clay. **Varved** – see Laminated. **Crumbly** – cohesive soils which break into small blocks or crumbs on drying. **Argillaceous** – having appreciable amounts of clay in the soil or rock mass. Used most often in describing limestones, occasionally sandstones. **Calcareous** – containing appreciable quantities of calcium carbonate. Can be either nodular or "powder." **Mottled** – characterized as having multiple colors organized in a marbled pattern. **Evaporite** – deposits of salts and other soluble compounds. Most commonly calcium carbonate or gypsum. May be in either "powder" or visible crystal form. **Ferruginous** – having deposits of iron or nodules, typically oxidized and dark red in color.	brn = brown dk = dark lt = light wx = weathered calc = calcareous sw = severely weathered cw = completely weathered n/a = not available b. = below **Engineering Units** pcf = pounds per cubic foot psf = pounds per square foot tsf = tons per square foot pF = picofarad psi = pounds per square foot ksf = thousand pounds per square foot kips = thousand pounds (force)	LL = Liquid Limit PL = Plastic Limit PI = Plasticity Index (LL-PL) γd = 95-Dry Unit Weight SPT = standard penetration test N = blows per foot from SPT SCR = standard core recovery RQD = rock quality designation RQI = see RQD qu = unconfined compressive strength

Terms Describing Consistency of Soil and Rock

COARSE GRAINED MATERIAL		SEDIMENTARY ROCK	
DESCRIPTIVE TERM	BLOWS/FT (SPT)	DESCRIPTIVE TERM	STRENGTH, TSF
very loose	0 – 4	soft	4 – 8
loose	4 – 10	medium	8 – 15
firm (medium)	10 – 30	hard	15 – 50
dense	30 – 50	very hard	over 50
very hard	over 50		

Describing Consistency of Fine Grained Soil

DESCRIPTIVE TERM	BLOWS/FT (SPT)	UNCONFINED COMPRESSION, TSF
very soft	< 2	< 0.25
soft	2 – 4	0.25 – 0.50
medium stiff	4 – 8	0.50 – 1.00
stiff	8 – 15	1.00 – 2.00
very stiff	15 – 30	2.00 – 4.00
hard	over 50	over 4.00

MLA LABS, INC.   2804 LONGHORN BLVD.   AUSTIN, TEXAS 78758   512/873-8899   FAX 512/835-5114

APPENDIX B

STANDARD FIELD AND LABORATORY PROCEDURES

STANDARD FIELD AND LABORATORY PROCEDURES

STANDARD FIELD PROCEDURES

Drilling and Sampling

Borings and test pits are typically staked in the field by the drillers, using simple taping or pacing procedures and locations are assumed to be accurate to within several feet. Unless noted otherwise, ground surface elevations (GSE) when shown on logs are estimated from topographic maps and are assumed to be accurate to within a foot. A Plan of Borings or Plan of Test Pits showing the boring locations and the proposed structures is provided in the Appendix.

A log of each boring or pit is prepared as drilling and sampling progressed. In the laboratory, the driller's classification and description is reviewed by a Geotechnical Engineer. Individual logs of each boring or pit are provided in the Appendix. Descriptive terms and symbols used on the logs are in accordance with the Unified Soil Classification System (ASTM D-2487). A reference key is also provided. The stratification of the subsurface material represents the soil conditions at the actual boring locations, and variations may occur between borings. Lines of demarcation represent the approximate boundary between the different material types, but the transition may be gradual.

A truck-mounted rotary drill rig utilizing rotary wash drilling or continuous flight hollow or solid stem auger procedures is used to advance the borings, unless otherwise noted. A backhoe provided by others is used to place test pits. Test pits are advanced to the required depth, refusal (typically bedrock) or to the limits of the equipment. Samples of soil are obtained from the borings or test pit spoils for subsequent laboratory study. Samples are sealed in plastic bags and marked as to depth and boring/pit locations in the field. Cores are wrapped in a polyethylene wrap to preserve field moisture conditions, placed in core boxes and marked as to depth and core runs. Unless notified to the contrary, samples and cores will be stored for 90 days, then discarded.

Standard Penetration Test and Split-Barrel Sampling of Soils (ASTM D-1586) (SPT)

This sampling method consists of driving a 2 inch outside diameter split barrel sampler using a 140 pound hammer freely falling through a distance of 30 inches. The sampler is first seated 6 inches into the material to be sampled and then driven an additional 12 inches. The number of blows required to drive the sampler the final 12 inches is known as the Standard Penetration Resistance. The results of the SPT is recorded on the boring logs as "N" values.

Thin-Walled Tube Sampling of Soils (ASTM D-1587) (Shelby Tube Sampling)

This method consists of pushing thin walled steel tubes, usually 3 inches in diameter, into the soils to be sampled using hydraulic pressure or other means. Cohesive soils are usually sampled in this manner and relatively undisturbed samples are recovered.

B-1

Soil Investigation and Sampling by Auger Borings (ASTM D-1452)

This method consists of auguring a hole and removing representative soil samples from the auger flight or bit at intervals or with each change in the substrata. Disturbed samples are obtained and this method is, therefore, limited to situations where it is satisfactory to determine the approximate subsurface profile and obtain samples suitable for Index Property testing.

Diamond Core Drilling for Site Investigation (ASTM D-2113)

This method consists of advancing a hole into hard strata by rotating a single or double tube core barrel equipped with a cutting bit. Diamond, tungsten carbide, or other cutting agents may be used for the bit. Wash water or air is used to remove the cuttings and to cool the bit. Normally, a 3 inch outside diameter by 2-1/8 inch inside diameter coring bit is used unless otherwise noted. The rock or hard material recovered within the core barrel is examined in the field and in the laboratory and the cores are stored in partitioned boxes. The intactness of all rock core specimens is evaluated in two ways. The first method is the Standard Core Recovery expressed as the length of the total core recovered divided by the length of the core run, expressed as a percentage:

$$SCR = \frac{\text{total core length recovered}}{\text{length of core run}} \times 100\%$$

This value is exhibited on the boring logs as the Standard Core Recovery (SCR).

The second procedure for evaluating the intactness of the rock cores is by Rock Quality Designation (RQD). The RQD provides an additional qualitative measure of soundness of the rock. This index is determined by measuring the intact recovered core unit which exceed four inches in length divided by the total length of the core run:

$$RQD = \frac{\text{all core lengths greater than 4"}}{\text{length of core run}} \times 100\%$$

The RQD is also expressed as a percentage and is shown on the boring logs.

Vane Shear Tests

In-situ vane shear tests may be used to determine the shear strength of soft to medium cohesive soil. This test consists of placing a four-bladed vane in the undisturbed soil and determining the torsional force applied at the ground surface required to cause the cylindrical perimeter surface of the vane to be sheared. The torsional force sufficient to cause shearing is converted to a unit of shearing resistance or cohesion of the soil surrounding the cylindrical surface.

B-2

THD Cone Penetrometer Test

The THD Cone Penetrometer Test is a standard field test to determine the relative density or consistency and load carrying capacity of foundation soils. This test is performed in much the same manner as the Standard Penetration Test described above. In this test, a 3 inch diameter penetrometer cone is used in place of a split-spoon sampler. This test calls for a 170-pound weight falling 24 inches. The actual test in hard materials consists of driving the penetrometer cone and accurately recording the inches of penetration for the first and second 50 blows for a total of 100 blows. These results are then correlated using a table of load capacity vs. number of inches penetrated per 100 blows.

Ground Water Observation

Ground moisture observations are made during the operations and are reported on the logs of boring or pit. Moisture condition of cuttings are noted, however, the use of water for circulation precludes direct observation of wet conditions. Water levels after completing the borings or pits are noted. Seasonal variations, temperatures and recent rainfall conditions may influence the levels of the ground water table and water may be present in excavations, even though not indicated on the logs.

STANDARD LABORATORY PROCEDURES

To adequately characterize the subsurface material at this site, some or all of the following laboratory tests are performed. The results of the actual tests performed are shown graphically on the Logs of Boring or Pit.

Moisture Content - ASTM D-2216

Natural moisture contents of the samples (based on dry weight of soil) are determined for selected samples at depths shown on the respective boring logs. These moisture contents are useful in delineating the depth of the zone of moisture change and as a gauge of correlation between the various index properties and the engineering properties of the soil. For example, the relationship between the plasticity index and moisture content is a source of information for the correlation of shear strength data.

Atterberg Limits - ASTM D-4318

The Atterberg Limits are the moisture contents at the time the soil meets certain arbitrarily defined tests. At the moisture content defined as the plastic limit, P_w, the soil is assumed to change from a semi-solid state to a plastic state. By the addition of more moisture, the soil may be brought up to the moisture content defined as the liquid limit, L_w, or that point where the soil changes from a plastic state to a liquid state. A soil existing at a moisture content between these two previously described states is said to be in a plastic state. The difference between the liquid limit, L_w, and the plastic limit, P_w, is termed the plasticity index, I_w. As the plasticity index

B-3

MLA LABS, INC.   2804 LONGHORN BLVD.   AUSTIN, TEXAS 78758   512/873-8899   FAX 512/835-5114

increases, the ability of a soil to attract water and remain in a plastic state increases. The Atterberg Limits that were determined are plotted on the appropriate log.

The Atterberg Limits are quite useful in soil exploration as an indexing parameter. Using the Atterberg Limits and grain size analysis, A. Casagrande developed the Unified Soils Classification System (USCS) which is widely used in the geotechnical engineering field. This system related the liquid limit to the plasticity index by dividing a classification chart into various zones according to degrees of plasticity of clays and silts. Although the Atterberg Limits are an indexing parameter, K. Terzaghi has related these limits to various engineering properties of a soil. Some of these relationships are as follows:

1. As the grain size of the soil decreases, the Atterberg Limits increase.
2. As the percent clay in the soil increases, the Atterberg Limits increase.
3. As the shear strength increases, the Atterberg Limits decrease.
4. As the compressibility of a soil increases, the Atterberg Limits increase.

Triaxial Shear Test - ASTM D-2850-70

Triaxial tests may be performed on samples that are approximately 2.83 inches in diameter, unless a smaller diameter sample was necessary to achieve a more favorable length:diameter (L:D) ratio. A minimum length to diameter ratio (L:D) of 2.0 is maintained to reduce end effects.

The triaxial tests are typically unconsolidated-undrained using nitrogen gas for chamber confining pressure. Confining pressures are selected to conform to in-situ hydrostatic pressure considering the earth to be a fluid of 120 pcf. In this test, undisturbed Shelby tube samples are trimmed so that their ends are square and then pressed in a triaxial compression machine. The load at which failure occurs is the compressive strength. The results of the triaxial tests and the correlated hand penetrometer strengths can be utilized to develop soil shear strength values.

Unconfined Compressive Strength of Rock Cores - ASTM D-2938

The unconfined compressive strength is a valuable parameter useful in the design of foundation footings. This value, qu, is related to the shearing resistance of the rock and thus to the capacity of the rock to support a load. In completing this test it is imperative that the length:diameter ratio of the core specimens are maintained at a minimum of 2:1. This ratio is set so that the shear plane will not extend through either of the end caps. If the ratio is less than 2.0 a correction is applied to the result.

Grain Size Analysis - ASTM D-421 and D-422

Grain size analysis tests are performed to determine the particle size and distribution of the samples tested. The grain size distribution of the soils coarser than the Standard Number 200 sieve is determined by passing the sample through a standard set of nested sieves, and the distribution of sizes smaller than the No. 200 sieve is determined by a sedimentation process,

B-4

using a hydrometer. The results are given on the log of Boring/Pit or on Grain Size Distribution semi-log graphs within the report.

Soil Suction Test - ASTM D-5298-94

Soil suction (potential) tests are performed to determine both the matric and total suction values for the samples tested. Soil suction measures the free energy of the pore water in a soil. In a practical sense, soil suction is an indication of the affinity of a given soil sample to retain water. Soil suction provides useful information on a variety of characteristics of the soil that are affected by the soil water including volume change, deformation, and strength.

Soil suction tests are performed using the filter paper method per ASTM D-5298. Results of these tests are shown graphically on the logs of boring and tabulated in summary sheet of laboratory data.

For matric suction values found using this method, it should be noted that when the soil is in a dry state adequate contact between the filter paper and the soil may not be possible. This lack of contact may result in the determination of total suction instead of matric suction.

B-5

METRIC UNITS, SYSTEM INTERNATIONAL (SI)

The most widely used system of units and measures around the world is the Systeme International d'Unites (SI), which is the modern form of the metric system. The other system of measurement is the U.S. Customary System Units, also known as "English Units," consisting of the mile, foot, inch, gallon, second, and pound. Although the English system is gradually being replaced by the metric system in some sectors of U.S. industry, the full conversion of the U.S. to the metric system is still incomplete.

Using metric units or providing metric equivalents is important because of the following reasons:

1. Use of metric units facilitates understanding and communication in technical areas such as engineering, architecture, building codes, and other scientific arenas at a global level.

2. Use of metric units is simpler because variations from smaller to larger units or vise versa are in multiples of 10, but in English units the multipliers could vary with unit and with subject. For example, smaller units of an inch are $\frac{1}{2}$, $\frac{1}{4}$, $\frac{1}{8}$, or $\frac{1}{16}$ of an inch, each a multiplier of 2 larger than the other. Conversion of inches to feet is at a multiplier of 12 and from feet to yards at a multiplier of 3. For metric, the smaller unit of a centimeter is a millimeter, which is a centimeter divided by 10. Conversion of centimeters to decimeters is by multiplying a centimeter by 10, and conversion of decimeters to meters is by multiplying by 10, and so on. Accordingly, computations and problem solving are prone to less error.

The conversion to metric units can take two forms, soft metrication and hard metrication.

Soft Metrication

Soft metrication is the use of metric units in specifying measurements, sizes, and other dimensions without changing product sizes and without changing the everyday practice of using English units. For example, a wood-stud member commonly used is a 2 × 4, which is actually 1.5 inches × 3.5 inches. To report or to specify the actual size of this member in metric, a soft conversion of 38 mm × 89 mm is used (rounded from actual 38.1 and 88.9). Another example could be the load-bearing pressure of clay soils of 1500 pounds per square-foot being reported as 72 KPa (kilo Pascal).

Hard Metrication

Hard metrication goes beyond soft metrication and converts production and manufacturing based on metric sizes. For example, instead of manufacturing 2 × 4 wood studs of 1.5 inches × 3.5 inches (38.1 mm ' 88.9 mm, actual dimensions), wood studs of 40 mm by 90 mm might be manufactured. Another example is $\frac{1}{2}$-inch diameter (12.7 mm) U.S. size automotive bolts versus 13 mm metric bolts, which are manufactured with a diameter of 13 mm. The production of other structural or nonstructural members such as structural steel, plywood, nails, pipes, ducts, insulation panels, and all other such elements would also be done in metric rather than manufacturing in English units and reporting metric equivalents.

More information on the SI system in the U.S. is available from the National Institute of Standards and Technology (NIST), an agency of the U.S. Department of Commerce:

http://physics.nist.gov/cuu/Units/index.html

or

http://physics.nist.gov/Pubs/SP811/contents.html

UNIT CONVERSION TABLES
SI SYMBOLS AND PREFIXES

BASE UNIT		
Quantity	Unit	Symbol
Length	Meter	m
Mass	Kilogram	kg
Time	Second	s
Electric curren	Ampere	A
Thermodynamic temperature	Kelvin	K
Amount of substance	Mole	mol
Luminous intensity	Candela	cd

SI SUPPLEMENTARY UNITS		
Quantity	Unit	Symbol
Plane angle	Radian	rad
Solid angle	Steradian	sr

SI PREFIXES		
Multiplication Factor	Prefix	Symbol
$1\ 000\ 000\ 000\ 000\ 000\ 000 = 10^{18}$	exa	E
$1\ 000\ 000\ 000\ 000\ 000 = 10^{16}$	peta	P
$1\ 000\ 000\ 000\ 000 = 10^{12}$	tera	T
$1\ 000\ 000\ 000 = 10^{9}$	giga	G
$1\ 000\ 000 = 10^{6}$	mega	M
$1\ 000 = 10^{3}$	kilo	k
$100 = 10^{2}$	hecto	h
$10 = 10^{1}$	deka	da
$0.1 = 10^{-1}$	deci	d
$0.01 = 10^{-2}$	centi	c
$0.001 = 10^{-3}$	milli	m
$0.000\ 001 = 10^{-6}$	micro	μ
$0.000\ 000\ 001 = 10^{-9}$	nano	n
$0.000\ 000\ 000\ 001 = 10^{-12}$	pico	p
$0.000\ 000\ 000\ 000\ 001 = 10^{-15}$	femto	f
$0.000\ 000\ 000\ 000\ 000\ 001 = 10^{-18}$	atto	a

SI DERIVED UNITS WITH SPECIAL NAMES			
QUANTITY	UNIT	SYMBOL	FORMULA
Frequency (of a periodic phenomenon)	hertz	Hz	1/sk
Force	newton	N	$g \bullet m/s2$
Pressure, stress	pascal	Pa	N/m2
Energy, work, quantity of heat	joule	J	N-m
Power, radiant flux	watt	W	J/s
Quantity of electricity, electric charge	coulomb	C	A-s
Electric potential, potential difference, electromotive force	volt	V	W/A
Capacitance	farad	F	C/V
Electric resistance	ohm	Ω	V/A
Conductance	siemens	S	A/V
Magnetic flux	weber	Wb	V-s
Magnetic flux density	tesla	T	Wb/m^2
Inductance	henry	H	Wb/A
Luminous flux	lumen	Im	$cd \bullet sr$
Luminance	lux	Ix	Im/m^2
Activity (of radionuclides)	becquerel	Bq	I/s
Absorbed dose	gray	Gy	J/kg

CONVERSION FACTORS

To convert	to	multiply by
LENGTH		
1 mile (U.S. statute) 1 yd 1 ft 1 in	km m m mm mm	1.609 344 0.9144 0.3048 304.8 25.4
AREA		
1 mile^2 (U.S. statute) 1 acre (U.S. survey) 1 yd^2 1 ft^2 1 in^2	km^2 ha m^2 m^2 m^2 mm^2	2.589 998 0.404 6873 4046.873 0.836 1274 0.092 903 04 645.16
VOLUME, MODULUS OF SECTION		
1 acre ft 1 yd^3 100 board ft 1 ft^3 1 in^3 1 barrel (42 U.S. gallons)	m^3 m^3 m^3 m^3 L(dm^3) mm^3 mL (cm^3) m^3	1233.489 0.764549 0.235 9737 0.028316 85 28.3168 16 387.06 16.3871 0.158 9873
(FLUID) CAPACITY		
1 gal (U.S. liquid)* 1 qt. (U.S. liquid) 1 pt. (U.S. liquid) 1 fl oz (U.S.) 1 gal. (U.S. liquid) *1 gallon (UK) approx. 1.2 gal (U.S.)	L** mL mL mL m^3 **1 liter approx. 0.001 cubic meter	3.785 412 946.3529 473.1765 29.5735 0.003 785 412
SECOND MOMENT OF AREA		
1 in^4	mm^4 m^4	416 231 4 416 231 4 x 10^{-7}
PLANE ANGLE		
1° (degree) 1′ (minute) 1″ (second)	rad mrad urad urad	0.017 453 29 17.453 29 290.8882 4.848 137

METRIC UNITS, SYSTEM INTERNATIONAL (SI)

To convert	to	multiply by
VELOCITY, SPEED		
1 ft/s	m/s	0.3048
1 mile/h	km/h	1.609 344
	m/s	0.447 04
VOLUME RATE OF FLOW		
1 ft^3/s	m^3/s	0.028 316 85
1 ft^3/min	L/s	0.471 9474
1 gal/min	L/s	0.063 0902
1 gal/min	m^3/min	0.0038
1 gal/h	mL/s	1.051 50
1 million gal/d	L/s	43.8126
1 acre ft/s	m^3/s	1233.49
TEMPERATURE INTERVAL		
1° F	°C or K	0.555 556
		5/9°C = 5/9K
EQUIVALENT TEMPERATURE (TOC = TK − 273.15)		
t_{oF}	t_{oC}	$t_{oF} = 9/5t_{oC} + 32$
MASS		
1 ton (short***)	metric ton	0.907 185
1 lb	kg	907.1847
1 oz	kg	0.453 5924
*** 1 long ton (2,240 lb)	g	28.349 52
	kg	1016.047
MASS PER UNIT AREA		
1 lb/ft^2	kg/m^2	4.882 428
1 oz/yd^2	g/m^2	33.905 75
1 oz/ft^2	g/m^2	305.1517
DENSITY (MASS PER UNIT VOLUME)		
1 lb/ft^3	kg/m^3	16.01846
1 lb/yd^3	kg/m^3	.593 2764
1 ton/yd^3	t/m^3	1.186 553
FORCE		
1 tonf (ton-force)	kN	8.896 44
1 kip (1,000 lbf)	kN	4.448 22
1 lbf (pound-force)	N	4.448 22
MOMENT OF FORCE, TORQUE		
1 lbf•ft	N•m	1.355 808
1 lbf•in	N•m	0.112 9848
1 tonf•ft	kN•m	2.711 64
1 kip•ft	kN•m	1.355 82

To convert	to	multiply by
FORCE PER UNIT LENGTH		
1 lbf/ft	N/m	14.5939
1 lbf/in	N/mk	175.1268
1 ton/ft	N/m	29.1878
PRESSURE, STRESS, MODULUS OF ELASTICITY (FORCE PER UNIT AREA) (1 PA = 1 N/M^2)		
1 tonf/in^2	Mpa	13.7895
1 tonf/ft^2	kPa	95.7605
1 kip/in^2	Mpa	6.894 757
1 lbf/in^2	kPa	6.894 757
1 lbf/ft^2	Pa	47.8803
Atmosphere	kPa	101.3250
1 inch mercury	kPa	3.376 85
1 foot (water column at 32° F)	kPa	2.988 98
WORK, ENERGY, HEAT (1J = 1N • M = 1W • WS)		
1 kWh (550 ft•lbf/s)	MJ	3.6
1 Btu (Int. Table)	kJ	1.055 056
1 ft•lbf	J	1055.056
	J	1.355 818
COEFFICIENT OF HEAT TRANSFER		
1 Btu/(ft^2•h•°F)	W/(m•K)	5.678 263
THERMAL CONDUCTIVITY		
1 Btu/(ft•h•°F)	W/(m^2•K)	1.730 735
ILLUMINANCE		
1 lm/ft2 (footcandle)	lx (lux)	10.763 91
LUMINANCE		
1 cd/ft^2	cd/m^2	10.7639
1 foot lambert	cd/m^2	3.426 259
1 lambert	kcd/m^2	3.183 099

INDEX

Updated Editions Revisit Critical Elements In IBC

SEISMIC AND WIND FORCES: STRUCTURAL DESIGN EXAMPLES, 4TH EDITION

This new edition has been updated to the 2012 *International Building Code®* (IBC®), ASCE/SEI 7-10, ACI 318-11, NDS-2012, AISC 341-10, AISC 358-10, AISC 360-10, and the 2011 MSJC Code. In each chapter, sections of the code are presented and explained in a logical and simple manner and followed by illustrative examples. Each example concentrates on a specific section of the code and provides a clear and concise interpretation of the code requirement. The text is organized into six chapters corresponding to the primary structural design sections of the code: Seismic design, Design for wind loads, Seismic design of steel structures, Seismic design of concrete structures, Seismic design of wood structures, and Seismic design of masonry structures.

More than 100 completely worked-out design examples are included. Problems are solved in a straight forward step-by-step fashion with extensive use of illustrations and load diagrams. (580 pages)

SOFT COVER #9185S4
PDF DOWNLOAD #8804P4

COLOR ILLUSTRATIONS!

A GUIDE TO THE 2012 IRC® WOOD WALL BRACING PROVISIONS

Wall bracing is one of the most critical, yet commonly misunderstood, safety elements in one- and two-family dwellings and town-houses constructed under the *International Residential Code®*. Wall bracing provisions can be complex because of a great number of aesthetic, cultural, economic and energy-related variables. 2012 IRC Section R602.12 contains a new Simplified Bracing Method that streamlines the procedure down to a five-minute process. This full-color illustrative guide provides examples and discussion of how to apply the new process, helping you save time and costly redesign. This guide makes provisions easy to understand by identifying and explaining the key elements of wall bracing provisions. It was developed by ICC and APA – The Engineered Wood Association to help building departments, designers, builders and others effectively apply the lateral bracing requirements of the 2012 IRC. (280 pages)

SOFT COVER #7102S12
PDF DOWNLOAD #8799P12

INCLUDES ONLINE BONUS ITEM!
COLOR ILLUSTRATIONS!

CONCRETE MANUAL: BASED ON 2012 IBC® AND ACI 318-11

The *Concrete Manual*, now updated to the 2012 *International Building Code®* (IBC) and ACI 318-11, provides the guidance and information that inspectors and other construction professionals need to become more proficient in concrete field practices and inspection.

The *Concrete Manual* will:

- Introduce concrete and explain what it is and why it behaves as it does
- Explain conventional concrete construction procedures
- Cover special concrete technologies such as Autoclaved, self-consolidating and pervious
- Discuss control and inspection procedures
- Explore statistical quality control methods and their application to concrete construction
- Detail proper field testing procedures
- Detail proper placement of reinforcement

A Resource Reference section includes a list of the concrete industry and technical organizations to contact for additional information. Your purchase of the *Concrete Manual* includes exclusive online access to the *Concrete Manual Workbook* to help you master concrete inspection and field practices. The workbook contains learning objectives, lesson notes, key points for studying, and quizzes for each chapter. The answer key includes references to the applicable sections in the *Concrete Manual*. (576 pages)

HARD COVER #9090S12
PDF DOWNLOAD #8853P12

REINFORCED MASONRY ENGINEERING HANDBOOK: CLAY AND CONCRETE MASONRY, 7TH EDITION

A new edition from Masonry Institute of America based on the requirements of the 2012 IBC that will eliminate repetitious calculations. The RMEH contains detailed explanations and applications of allowable stress design and strength design procedures, and more than 80 step-by-step examples.

HARD COVER #9346S7

12-07184